Building Contractor's Checklists and Forms

Building Contractor's Checklists and Forms

Sidney M. Levy

McGraw-Hill

New York Chicago San Francisco Lisbon London Madrid
Mexico City Milan New Delhi San Juan Seoul
Singapore Sydney Toronto

The McGraw-Hill Companies

1 2 3 4 5 6 7 8 9 0 QPD/QPD 0 1 0 9 8 7 6 5 4

ISBN 0-07-144172-7

The sponsoring editor for this book was Larry S. Hager and the production supervisor was Pamela A. Pelton. It was set in Weiss by Lone Wolf Enterprises, Ltd. The art director for the cover was Anthony Landi.

Printed and bound by Quebecor/Dubuque.

About the Author

Sidney M. Levy is an independent construction industry consultant with more than 30 years' experience. He is the author of numerous books on construction methods and operations. He received the British Chartered Institute of Building Silver Medal for the third edition of *Project Management in Construction*, in the Managing Construction category. He lives in Baltimore, Maryland.

Table of Contents

Section 3
Checklists for Residential Contractors. 93

Section 4
Subcontractor Interview Forms 105

Section 5
Building Contractor's Comprehensive Checklist 157

Section 6
Useful Forms for Builders. 195

Preface

We all strive to build the highest quality construction projects and the subject of Quality Control and Quality Assurance always becomes a hot topic. What standards or goals do we anticipate achieving? How do we define quality and, as important, how can we best implement a plan or program to meet these standards or goals?

Well, first of all, let's differentiate between Quality Control (QC) and Quality Assurance (QA). There is often confusion on the jobsite as to what we mean by these terms and they are sometimes thought to be interchangeable—but they are not. Quality Control is best defined as the standard established during the design of a construction component, assembly, or equipment used in that component or assembly as incorporated in the "contract" plans and specifications. Quality Assurance is the activity taken to "assure" that the quality standards contained in the contract documents have been met.

As an example, let's say the architect, when designing a cast-in-place concrete column, uses as the quality standard they wish to achieve, specifications established by the American Concrete Institute (ACI). ACI has quality standards for cast-in-place and precast concrete. The dimensional tolerances for concrete walls, for example, in their Specification 301, 2.1.1.1 is as follows. In lines and surfaces of columns, piers, walls:

- Plumb in any 10 feet +/- ¼ inch

- Maximum for total height of the structure (taken to be less than 100 ft) +/- 1"

So that would be the QC standard that should be followed in both forming and pouring a concrete column designed to meet ACI requirements. Upon completion of the column, an inspection check is made of the structure to determine (assure) that these quality requirements have been met. Placing a level on the column will determine if it meets a vertical plane within the specified ¼ inch in 10 feet of height—that's Quality Assurance (QA).

The father of Quality Control was J. Edwards Deming; his philosophy was—that which can be measured can be controlled. In the 1950s, having been brushed aside by the automobile industry in this country, he traveled to Japan hoping to interest the Japanese car makers in ways in which QC could be applied to that industry. His ideas were eagerly embraced by what is now Toyota and Nissan—and the rest is history. His maxim of "what can be measured can be controlled" is just as applicable today as it was 50 years ago.

Another term you'll hear today is Total Quality Control (TQC) which involves not only design and construction, but all other elements of the construction process: training of personnel, shifting the responsibility for detecting defects from inspectors to workers, working with material suppliers to increase the quality of their products, and improving the performance of every department within the construction company's organization.

The Inspection Process

The ultimate goal of an inspection program is to place first-line responsibility for quality on the workers actually engaged in the construction process, having the construction project manager or supervisor perform a formal inspection to insure that quality levels have been achieved. Inspections are required at various stages in the construction process to insure that a previously completed component or assembly will allow work to follow and continue the links in the quality chain.

If the steel or wood framing of a partition wall is not true in both vertical and horizontal planes, obviously the application of gypsum drywall that follows will not be true in either of these two planes. So an inspection of framing (QA) to determine if the design standards (QC) have been met is necessary to insure the quality of that component and the process that follows—be it painting or wall covering. A defect caught early on reduces the cost and time required for correction—another reason for a series of inspections at the jobsite, particularly in today's environment where time of construction is being compressed more and more.

Inspection checklists can be of great assistance in assuring that construction components or assemblies meet or conform to pre-determined standards. If need be, these forms can actually include some of the quality standards to help the inspectors in their task. Contract specifications, for concrete as an example, often include a Tolerance paragraph, but they may only include "Conform to ACI 301, 4.3" assuming that every contractor has all four volumes of the ACI Manuals on the job site. In those instances, it would be helpful to contact the architect, find out the actual tolerances required to meet ACI 301,4.3, and include them on the inspection form.

This book contains several types of "checklists":

- Checklists to assist in performing various project management tasks such as Job Site Mobilization, Submittal Review, Winter Conditions, and Safety.
- Trade Inspection Checklists ranging from Earthwork to Electrical.
- Several checklists relating specifically to residential construction, such as a Rough Trades Inspection and a Pre-Inspection form for owner occupancy.
- Subcontractor Interview checklists to be prepared prior to a proposal review meeting with the subcontractor and listing specific items to be addressed to accurately define scope and price.
- Building Contractor's comprehensive checklists.

Used as a Memory Jogger or as training and instruction aids, these checklists can be customized to fit specific projects or specific work tasks and they will probably permit a supervisor to conduct an inspection more quickly and more thoroughly.

But checklists, by themselves, are not the cure-all for quality problems. If not properly used they can be a waste of everyone's time. Checking boxes quickly and turning them over to a subcontractor, or the company's foreman for correction or re-work is only part of the story. A checklist should also be used to highlight the nature of a problem and whether it is an isolated one or a recurring one. A checklist also serves to identify trouble spots, allowing a supervisor to trace them back to their source, and correct them at that source.

Does it make much sense to inspect framing in a number of rooms only to find the same problems repeating themselves? Instead of merely having the tradesman correct them, it would be more productive to meet with your foreman or the subcontractor's foreman to discuss these issues and determine their cause—poor supervision, inadequately trained mechanics, poor quality materials, disgruntled workers? A little time spent investigating the matter at this stage will be beneficial in uncovering why sub-standard work is being performed, what is required to correct it and hopefully prevent it from spreading—making everyone's life less stressful.

Quality needs to become foremost in the minds of all project managers, project superintendents, subcontractors and their workers in order for the Checklist program to succeed. Don't let it morph into a paper blizzard that becomes the latest coffee break joke.

Section 1

Checklists for Construction Component Field Inspections

Construction Component Inspections

Component or assembly inspections included in this section may be required for many purposes:

1. To insure that the work is performed as previously agreed upon or contracted for
2. To insure that the work meets the contract quality standards
3. To insure that the work is in compliance with the contract documents
4. To verify that the work has been completed according to the time and sequence indicated in the construction schedule
5. Establishing areas requiring re-work
6. Providing documentation for back-charges
7. Preparing a Punch List
8. Checking status of work to justify payment to a subcontractor
9. Provide documentation for project close-outs
10. Provide inspection reports for use in project meetings

The following checklists are those most commonly used in commercial and institutional projects today. Of course, each checklist can be modified to conform to a specific work task or an unusual one.

Inspection Checklist

Acoustical Treatment **Project No.__________**

1. Shop drawings, product data, samples are approved and on site._________
2. Verify that contractor understands layout of suspension system and has coordinated work with other trades i.e. HVAC, Electrical. _________
3. Hangers are of proper size, material, gauge, spacing. _________
4. Intermediate hangers, if field conditions require, are installed. _________
5. Hangers anchored to concrete support are installed as required and are fastened by twisting three times. _________
6. If anchored by powder-actuated anchors, a minimum of three twists are required. _________
7. If anchored by expansion shields drilled into concrete, minimum of three twists are required. _________
8. Hangers anchored in steel support systems are installed as required. _________
9. Isolators are provided and installed in accordance with manufacturer's instructions. _________
10. Turnbuckles are provided, if required. _________
11. Sway bracing is installed, if required. _________
12. Suspension system, components, and accessories are provided and installed as approved and required. _________
13. Perimeter and edge conditions are provided and installed as required. _________
14. Joint treatment is consistent. _________
15. System is installed and adjusted in true alignment, even and level. _________
16. Fire-rated systems are installed in accordance with UL requirements and include hold down clips. _________
17. Access to equipment above ceiling is provided as required. _________
18. Identification of access is provided by use of labels on grid, colored pins. _________
19. Sound isolation elements above ceiling provided and installed as required. _________
20. Exposed rivets are painted. _________

General Notes:____

____

____

____

Inspected by:________________________Date:______________

Figure 1.1 Acoustical Treatment Checklist.

Inspection Checklist

Acoustical Tile, Board, Insulation, Barriers Project No.________

Tile and Board

1. Material is of type, material, pattern, and edge condition required._______

2. Material is stored in climatized area. _______

3. Material is installed in accordance with mounting required. _______

4. Adhesives and air space are in accordance with manufacturer's instructions_______

5. If tile is to be sprayed or touched-up, verify paint material is approved for application and will not affect the acoustical performance. _______

Acoustical Insulation and Barriers

1. Batts are of thickness and density required and are identified. _______

2. Batts are installed and secured tightly to all adjoining surfaces, cut-outs, edges, etc. and all spaces are completely filled. _______

3. Suspended insulation is well-secured, tight-fitting, and sealed as required. _______

4. Wall board insulation is provided and installed and in location as required before the installation of finish surfaces. _______

General Notes:____

____

____

____

____

____

____

Inspected by:______________________________Date:____________________

Figure 1.2 Acoustical Tile, Board, Insulation, Barriers Checklist.

Inspection Checklist

Base and Cap Flashings **Project No.__________**

1. Flashing is provided to suit conditions- cant, size, gauge, fabrication. _________
2. Base flashing extends up sufficiently. _________
3. Flange is embedded at least 4" in roof membrane. Method of embedment is per manufacturer's approved product data. (It is good practice to cover as much metal as practical to avoid movement from temperature variations.) _________
4. Seams are lapped, locked and soldered, welded, riveted as required. _________
5. Secure anchorage is provided for size, spacing, and fixing of cleats or other equipment mountings. _________
6. Cap flashings are of shapes, sizes and gauges required and are installed to provide secure anchorage, allow movement, have sufficient laps and spacing. _________
7. Counter flashing is extended sufficiently into masonry walls or into reglet and is securely anchored and caulked, if necessary. _________
8. Reglets are provided at required areas; observe setting in concrete or masonry to insure firm anchorage. _________
9. Reglets are protected to prevent deformation or filling during installation. _________
10. Observe installation of sheet metal into reglets for tightness, weatherproofing, caulking, and lap. _________
11. Plastic flashing is to type required and is installed in accordance with the manufacturer's recommendations. _________

Wall and Thru Wall Flashings:

1. Verify that the contractor understands the location for flashing fabrication and design. _________
2. Lap, turn up, location in wall, depth of masonry, and length are as required. _________
3. Sill flashing and pans extend full depth, are turned up, extend at least 4", and are installed for drainage. _________
4. End dams are installed if required. _________

General Notes:____

Inspected by:____________________Date:________________

Figure 1.3 Base and Cap Flashings Checklist.

Inspection Checklist

Concrete Reinforcement **Project No.**____________

1. Shop drawings are approved and on site._________
2. Grade of steel delivered as required. _________
3. Spacing coordinated to suit masonry/concrete units. _________
4. Required clearance of steel from forms provided. _________
5. Length of splices and staggered splices as required. _________
6. Bends within radii and tolerance are uniformly made. _________
7. Additional bars at intersections, openings, and corners provided. _________
8. Bars cleaned of materials that affect bond. _________
9. Dowels for marginal bars at openings. _________
10. Bars tied and supported to avoid displacement. _________
11. Spacers, tie wires, chairs as required. _________
12. Conduit is separated by 3 conduit diameters minimum. _________
13. No conduit or pipe placed below rebar material except where approved. _________
14. No contact of bars is made with dissimilar metals. _________
15. Bar not near surface which may cause rusting. _________
16. Adequate clearance provided for deposit of concrete. _________
17. Verify that contractor has resolved conflicts with embeds. _________
18. Verify that contractor has coordinated for anchors, piping, sleeves. _________
19. Special coating as required. _________
20. No bent bars and tension members installed except where approved. _________
21. Unless approved, boxing out is not approved for subsequent grouting out. _________

24. Rules for bar splices:
    For 24d lap – multiply bar size by 3 = lap in inches
    For 32d lap – multiply bar size by 4 = lap in inches
    For 40d alp – multiply bar size by 5 = lap in inches

25. Agency/Engineer inspection is performed, if required. _________

General Notes: ____

Inspected by:____________________Date:____________

Figure 1.4 Concrete Reinforcement Checklist.

Inspection Checklist

Concrete Placement **Project No.**____________

1. Shop drawings approved and on site. ________
2. Verify correct psi ordered from plant. ________
3. Chutes, elephant trunks required? ________
4. Verify approval of forms and rebar prior to pour. ________
5. Requirements for testing, mix design, ingredients. ________
6. Test lab notified and tests required. ________
   - Slump ________
   - Number of cylinders ________
   - Temperature/truck waiting time ________
7. Testing required at plant. ________
8. Vibrators to be used during pour. ________
9. Temporary form openings O.K.? ________
10. Arrange for specified curing and saw cut joints. ________
11. Arrange for cold weather protection. ________

    -or
12. Arrange for hot weather protection. ________
13. Embeds available for insertion in pour. ________
14. Box-out properly installed in form work. ________
15. Verify finishes – smooth troweled, broom. ________
16. No troweling while bleed water is on surface. ________
17. Slopes to drain properly designated. ________
18. Wet spray or curing compound adequately performed. ________
19. Traffic over area controlled. ________
20. Preparations for repairs at hand. ________

General Notes:________________________

Inspected by:________________ Date:____________

Figure 1.5 Concrete Placement Checklist.

Inspection Checklist

Concrete Form Removal

Project No.__________

1. Method of patching approved and applied early.__________
2. No troweling while white bleed water is on surface. __________
3. Overtroweling is to be avoided. __________
4. Finishing method provides surface within specified tolerances. __________
5. Slopes provide proper drainage. __________
6. Check edges that are finished and match. __________
7. Moist curing is adequately performed. __________
8. Proper covering of slabs is provided. __________
9. Curing compounds are applied per specification. __________
10. Saw cuts performed as required. __________
11. Zip strip flanges removed. __________

General Notes:__________

Inspected by:__________Date:__________

Figure 1.6 Concrete Form Removal Checklist.

Inspection Checklist

Electrical – Grounding **Project No.____________**

1. Visually observe all grounding system conductors, connections, and electrodes as work progresses. __________

2. For water piping system used, check that pipe is metallic and that no insulating fitting is interposed in pipe between ground wire connection point and interior or exterior pipe system. __________

3. Contact surfaces are clean and dry, are metal-to-metal, and tight bolt connections are made. __________

4. Observe size, length, number, material, and installation of ground rods. __________

5. Ground conductor in all conduits. __________

General Notes: ____

Inspected by:____________________Date:______________

Figure 1.7 Electrical - Grounding Checklist.

Inspection Checklist

Electrical - Lighting **Project No.**__________

1. Approved shop drawings on site. __________

2. Observe lighting layout is coordinated with architectural drawings and any discrepancies are reported. __________

3. Observe lighting layout is coordinated with plumbing, HVAC, fire protection drawings and any discrepancies are reported. __________

4. Suspension, supporting, and mounting methods are as required. NEC requires different method based upon weight of fixture. __________

5. Observe plumbness and alignment. __________

6. Observe mounting height and location as required. Fixtures should provide at least 6' 8" in head clearance. __________

7. Lamp type as required - wattage, energy saving, style, color, characteristics, i.e. long life. __________

8. Note fluorescent lamps at start of installation - CW, DL, WW, WWX etc. Check that lamps are operating properly. __________

9. Lamps are new and installed before completion, or re-installed if required. __________

10. Frames and accessories are as required - compatible with adjacent surfaces, no light leaks, weatherproof or corrosion resistant, finishes match, etc. __________

11. Check layout in mechanical spaces. __________

12. Check specifications for spare parts, attic stock. __________

General Notes:__________

Inspected by:__________ Date:__________

Figure 1.8 Electrical - Lighting Checklist.

Inspection Checklist

Electrical - Panelboards **Project No:__________**

1. Equipment is stored properly. __________

2. Boards are rigidly and securely mounted to floors and walls with attachments of sufficient strength to resist lateral forces. __________

3. Spaces are provided as required by contract documents for future circuits. __________

4. Spare breakers are provided and installed per contract documents. __________

5. Location and isolation of boards has been coordinated by the contractor with other trades. Working space required by code has been provided. __________

6. Panel cover closes properly and securely. __________

7. Keys provided. __________

8. Circuit directories typed and inserted. __________

General Notes:____

Inspected by:____________________Date:________________

Figure 1.9 Electrical - Panelboards Checklist.

Inspection Checklist

Electrical – Main Switch, Switchboard **Project No:**__________

1. Approved shop drawings on site. __________
2. Equipment properly stored. __________
3. Housekeeping pads are installed as required. __________
4. Equipment rigidly bolted to floor. __________
5. Spares and spaces per contract documents. __________
6. Code clearance at front of gear. __________

General Notes:__________

Inspected by:__________ Date:__________

Figure 1.10 Electrical - Main Switch, Switchboard Checklist.

Inspection Checklist

Electrical – Raceways **Project No:**__________

1. Approved shop drawings on site. __________
2. Equipment stored properly. __________
3. Observe limitations required on use of rigid conduit, thinwall, flexible metal conduit, plastic conduit, liquidtight metal conduit, etc. __________
4. Conduit size as required. __________
5 Required types of fittings are provided and are compatible with conduit type. __________
6. Conduits are run to all locations required and properly installed before slabs on grade are installed. __________
7. Stub-ups, couplings, etc. are installed above or at finish floor level for free-standing equipment as required. __________
8. Exposed conduit should be installed so that bent portion will not extend above the floor level. __________
9. Type of fittings as required and as approved - concrete tight, rain tight, cast, etc. Check for tightness to maintain ground continuity. __________
10. Conduit is secured and fastened as required and in accordance with applicable code(s). __________
11. Runs in wet areas are spaced at least ¼" off surfaces. __________
12. Verify areas where exposed conduit is allowed and conditions are understood. __________
13. Exposed conduits are installed parallel or perpendicular to structure. Vertical runs are plumb. __________
14. Raceways are kept closed during construction. __________
15. Coating and surface treatment are provided as required, including connections. No treatment is normally provided if enclosed in concrete. __________
16. Depth of installation in relation to finish grade is as required. Concrete or sand encasement is provided as required. Backfill as required. __________
17. Pull wires are provided, extend full length, and are of type required. __________
18. Stub-ups for future extensions are provided as required. Location identification is provided and recorded. __________
19. Means are provided to accommodate contraction and expansion at building expansion joints, as required. __________

General Notes:__________

Inspected by:__________ Date:__________

Figure 1.11 Electrical - Raceways Checklist.

Inspection Checklist

Electrical – Wire and Cable **Project No.** ____________

1. Materials are properly stored. ____________
2. Type of conductor material, size, stranding, and insulation is as required. ____________
3. Pulling of conductors and cables is as required using suitable equipment and methods. Allow no damage to sheath jackets or insulation. ____________
4. Color coding of wires where required. Use only white or natural grey identified conductors for Neutral. ____________
5. Neutral must be insulated throughout unless excepted by code. ____________
6. Grounding conductor for equipment when run with circuit conductors is to be bare or green color. ____________
7. Pipe tracing properly installed and labeled. ____________

General Notes: ____________

Inspected by: ____________ Date: ____________

Figure 1.12 Electrical - Wire and Cable Checklist.

Inspection Checklist

Electrical - Wiring Devices **Project No.** ____________

1. Approved shop drawings on site. ____________
2. Materials stored properly. ____________
3. Architectural drawings are referred to for comparison of all conditions affecting layout of outlets. Observe that the contractor has coordinated work with other trades. ____________
4. Wall receptacles, switch outlets, and fixture outlets are mounted at height and location required. ____________
5. Junction, pull, and outlet boxes are of type, size, and location required.____________
6. Boxes are securely and rigidly supported and do not rely on conduits for this support, except as permitted by code. ____________
7. Boxes are accessible. ____________
8. Do not allow excessive number of conductors in boxes, to be per code.____________
9. Unused openings to be closed.____________
10. Grounding continuity is maintained, including jumper, if required.____________
11. Installed devices are of required type, voltage, amperage, color, etc.____________
12. Device plates are of material, type, ganging, finish as required.____________
13. Pilot lights are provided as required.____________
14. Plates cover openings completely and are in contact with finish material.____________
15. Plates are plumb and not dished or bowed.____________
16. Surface mounted boxes are provided with compatible plates and without overhang edges.____________
17. Neutral of multi-wire circuit will not be interrupted by removal of the device or fixture.____________
18. Weatherproof boxes where required.____________

General Notes:____________

Inspected by:____________Date:____________

Figure 1.13 Electrical - Wiring Devices Checklist.

Inspection Checklist

Electrical - Motors and Starters

Project No. __________

1. Motors have voltage rating and number of phases to suit supply system. __________
2. Motor subject to vibration or mounted on adjustable bases are connected with flexible metal conduit. __________
3. Liquid tight or explosion-proof flexible metal conduit provided as required. __________
4. Flexible metal conduit length is provided for each motor and motor starter as required. __________
5. Manual disconnect switch is provided for each motor and motor starter as required. __________
6. All control accessories are furnished as required: start-stop buttons, pilot lights, selector switches, and similar devices. __________
7. Motor nameplate full-load rated currents are compared with rating of motor running overcurrent protective devices (heaters). __________
8. Heaters of proper size are installed in starters. __________
9. Motor controllers are as required. Observe that horsepower and voltage rating is to be at least equal to motor controlled. __________
10. Each controller with disconnect complying with code. Check for 2 speed. __________
11. Nameplates are provided. __________

General Notes: ______________________________

Inspected by: ______________ Date: ______________

Figure 1.14 Electrical - Motors and Starters Checklist.

Inspection Checklist

Entrance Mats **Project No:__________**

1. Approved submittals, shop drawings, samples, product data on site. __________
2. Materials are properly stored on site and protected. __________
3. All material furnished and of approved type and color. __________
4. All accessory items furnished and of approved type. __________
5. Observe coordination and scheduling with other trades. __________
6. If recessed, of proper dimension and depth. __________
7. Entrance mats are suitably and securely attached. __________
8. Protective coating applied as required to surfaces of entrance mat frames. __________
9. Grilles installed with upper edge level with finish floor. __________
10. Adequate clearance for door movement. __________

General Notes:____

Inspected by:____________________Date:_______________

Figure 1.15 Entrance Mats Checklist.

Inspection Checklist

Finish Carpentry **Project No.**__________

1. Shop drawings and samples approved and on job. __________
2. Furring and blocking have been installed to receive finish materials. __________
3. Certificates of grade stamp are provided for all lumber. __________
4. Materials are suitably stored to prevent moisture or physical damage. __________
5. Materials have adequate temporary bracing, skids, etc. to prevent wracking, loosened members, or other defects due to handling. __________
6. Substrate and finishes are as required, visually inspected for evenness. __________
7. Method of attachment is as required. __________
8. Verify that contractor has coordinated with other trades. __________
9. Accessories such as scribe and trim molds are included. __________
10. Installed materials are to be adequately protected. __________
11. Tops are provided as required. __________
12. Cutting of holes for sinks and other appliances have been made. Check templates if available. __________
13. Surfaces are thoroughly cleaned, except for user's instructions. __________

General Notes:__________

Inspected by:__________ Date:__________

Figure 1.16 Finish Carpentry Checklist.

Inspection Checklist

Finish Grading **Project No._______________**

1. Subgrade is scarified and prepared to receive finish grading. _________
2. Stockpile topsoil is distributed. _________
3. Topsoil tested/modified to meet specifications. _________
4. Import approved topsoil, as required. _________
5. Observe installation of sleeves, raceways, boxes, piping required for irrigation and drainage. Electrical is provided as required. _________
6. Installation and location of all site improvements are coordinated. _________
7. Areas to receive plantings have not been excessively compacted. _________
8. Scheduling of site improvements, plantings to prevent damage. _________

General Notes:____

Inspected by:_________________________Date:_______________

Figure 1.17 Finish Grading Checklist.

Inspection Checklist

Finish Hardware - General and Butts and Hinges Project No.________

1. Hardware schedule, product data, samples, approved and on site. __________
2. Hardware installed in accordance with manufacturer's templates. __________
3. Finishes are as required and finishes match in each area. __________
4. Hardware is removed or protected during painting operations. __________
5. Recommended order of inspection:

   In hardware storage area before installation __________

   Door butts and hinges during and after installation __________

   Locksets, latchsets, and exit bolts during and after installation __________

   Door closers, after installation __________

   Door stops, holders, push-pulls, kickplates, after installation __________

Butts and Hinges

6. Ball bearing, iolite or nylon type, is provided as required. __________
7. Solid brass, bronze, aluminum, or stainless steel is provided as required. __________
8. Fire door hinges are steel with ball bearings or as otherwise approved for a labeled assembly. __________
9. Mortise type hinges are mortised flush. __________
10. Mortise hinges on door leaf to ¼" from stop side of door and jamb leaf 5/16" from stop (3/8" and 7/16" on very thick doors) unless otherwise noted. __________
11. Unless other noted, top hinge is mounted 5" below finish door frame and bottom hinge is mounted 10" above finish floor. Intermediate hinges are spaced and mounted equidistant from top and bottom hinges and from each other. __________
12. Sufficient throw is provided to clear trim and leaf can swing functionally as required. __________
13. One-half surface hinges are used on composite doors. __________

General Notes:____

Inspected by:____________________Date:______________

Figure 1.18 Finish Hardware - General and Butts and Hinges Checklist.

Inspection Checklist

Finish Hardware – Stops, Holders, Plates, Miscellaneous Items **Project No.**__________

1. Provide every door with a stop as required by hardware schedule. __________

2. Stops or holders to be attached to wallboard, plaster, etc. are screwed to solid blocking. __________

3. Verify wiring, outlet boxes, etc. are provided for magnetic holders. __________

4. Magnetic holders to be installed horizontally in same location as closer to prevent door warp unless otherwise noted. __________

5. Centerline of push plate is 45" from finish floor and centerline of pull plate is 42" from finish floor, unless otherwise noted. __________

6. Kick plates clear stops on push side of door and not more than 1" is exposed on each edge of door unless otherwise noted. __________

7. Push, pull, and kickplates are attached per specifications and manufacturer's recommendations. __________

Miscellaneous Items

8. Sliding door tracks are installed level and door is plumb. If separate tracks are used, bracket supports are to be directly over hangers when door is open or closed (especially required on fire-rated assemblies). Spacing and number of brackets are provided as required. __________

9. Thresholds are of required size, type, and interlock and are anchored as required. __________

10. Weather stripping and sound stripping allow for proper operation of door. __________

11. All hardware is complete and with required type and number of bolts, screws, and fastening devices installed. __________

12. Keying instructions are understood and keys delivered to Owner as required. Observe that all construction locks are removed and permanent cores are provided. __________

General Notes: ____________________

____

____

____

____

Inspected by:____________________Date:__________

Figure 1.19 Finish Hardware - Stops, Holders, Plates, Miscellaneous Items Checklist.

Inspection Checklist

Flashing and Sheet Metal **Project No.__________**

1. Approved shop drawings, product data, samples on site. __________
2. Delivered material is of approved type, shape, gauge, metal, fabrication, color, priming as required and all accessories are provided. __________
3. Isolation provisions are made for dissimiliar metals. Do not allow copper and aluminum to be in contact with each other or with ferrous metals. __________
4. Copper and aluminum flashing to be fastened with non-ferrous screws or nails and ferrous equipment bases not to be set on copper flashings. __________
5. Expansion joints are provided and installed as required or as specified. __________
6. Note location of expansion joints with respect to drains, downspouts, scuppers, corners, and other outlets. __________
7. Observe methods of installation - nailing and cleating types for spacing and location, also soldering, welding, bolting, riveting. __________
8. Flashing does not interfere with structural members. __________
9. Edge metal lapped a minimum of 4" with 12" staggered fastening through the back flange, unless otherwise required. __________
10. All edge metal laps are coated with plastic cement or manufacturer's approved materials on horizontal flange and vertical rise. Coating to cover entire lap and is sandwiched between. __________
11. Lengths are as long as practical, but generally not over 20 feet. __________
12. Installation coordinated with roofing installation. __________
13. Nailer or cant strip is provided for fastening flashing to roof deck and is of proper size, material, shape. Well secured and allows for venting if required. __________
14. Inner flange is applied over felt, lapped with polyisobutylene tape and properly nailed, when required. __________
15. Flashing is embedded in the roof membrane assembly and additional strip piles of membrane are provided as required by roofing system. __________
16. Method of anchoring lower edge of fascia is as required. Observe alignment, stiffness, etc. __________
16. Gravel stops are to be flush with deck unless otherwise noted. __________
17. Expansion joints, concealed or standing, are provided midway between outlets or as required. __________
18. Scuppers are installed low enough not to dam water on roof. __________
19. Overflow drains/scuppers are located properly at low points. __________
20. Accessories provided if required - basket strainer, bird screens, covers, etc. __________

General Notes:__________

Inspected by:__________ Date:__________

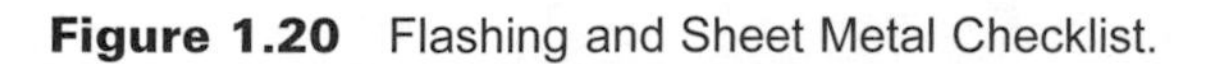

Figure 1.20 Flashing and Sheet Metal Checklist.

Inspection Checklist

Hollow Metal Doors and Frames **Project No.____________**

1. Shop drawings and schedule are approved and on site. ____________
2. Doors are as approved: type, design, material, accessories, etc. ____________
3. Check panel, lights, louvers, other features. ____________
4. Check for defects: dents, buckles, wraps. ____________
5. Check fabrication for construction and workmanship. ____________
6. Smooth edges, joints, finish, and straightness. ____________
7. Additional reinforcement as required for hardware. ____________
8. Observe backing plates during drilling operations. ____________
9. Observe that closure channels are provided. ____________
10. Provisions to receive hardware are adequate. ____________
11. Backset matches with hardware requirements. ____________
12. Stile edges, astragals required for pairs of doors. ____________
13. Fire-rated doors have labels and proper identification. ____________
14. Fire-rated frames have labels and proper identification. ____________
15. Fabrication and construction of frames as required. ____________
16. Frames are pre-braced as required. ____________
17. Extra reinforcement on frames at head, corners, and hardware. ____________
18. Proper type and number of anchors are provided. ____________
19. Verify adequate anchorage made during installation. ____________
20. Sound deadening treatment provided, if required. ____________
21. Provide features such as silencer holes. ____________
22. Frame is grouted during installation, if required. ____________
23. Frame is caulked, if required. ____________
24. Frames adequately braced where "built-in". ____________
25. Provide spreaders when installed in masonry walls. ____________
26. Protect threads of hinge plates in buck. ____________
27. Fusible link holders provided, if required. ____________
28. Observe installation of door and proper clearances. ____________
29. Doors are hung straight, level, and plumb. ____________
30. Door functions smoothly and easily. ____________
31. Hardware is properly adjusted. ____________
32. Observe glazing operation. ____________
33. Wire glass is provided, if required. ____________
34. Factory prime is touched up. ____________
35. Surfaces are adequate to receive finish. ____________
36. Report doors that cannot be properly cleaned. ____________
37. Bumper buttons installed. ____________
38. Protection provided to avoid marring. ____________

General Notes: ____

____

____

____

Inspected by:________________________Date:________________

Figure 1.21 Hollow Metal Doors and Frames Checklist.

Hydrostatic Test Report

Project:______________________ Project No.______________

Date:_________________ Test Report No.___________

System:____________________ Spec. Section:______________

System Pressure Test:_______psi

Pressure Gauge Reading:

Location:___________________ Start:__________

Start Time:__________ Finish:_________

Finish Time:_________

Comments:____

____

____

____

____

____

____

Test Accepted: Yes ___ No___

Re-test Required Yes___ No___

Witnessed By:

Contractor:__________________________________ Date:___________

Engineer:___________________________________ Date:___________

Owner's Representative_______________________ Date:___________

Note: This pressure test report must be filled out every time a test is performed.

1 copy to Contractor

1 copy to Engineer

1 copy to Owner Rep.

Figure 1.22 Hydrostatic Test Report Checklist.

Inspection Checklist

Hydraulic Elevators **Project No.______________**

1. Trade qualifications received, approved, and are on site. ______________
2. All permits, inspections, and tests obtained and documents and reports are on the site. ______________
3. Approved submittals, shop drawings, samples, product data, certificates, calculations, as required, are on the site. ______________
4. All wiring diagrams, manuals, instructions, as required, are received. ______________
5. Pre-construction conference held with appropriate trades. ______________
6. Materials are properly stored on site and protected. ______________
7. All materials and equipment furnished and of approved types. ______________
8. All accessory items furnished and of approved types. ______________
9. All car fixtures and equipment furnished and of approved types. ______________
10. Flooring material approved and contractor assigned. ______________
11. Observe coordination with other trades. ______________
12. All supporting elements have suitable structural attachments at shaftways. __________
13. Check anchoring devices and inserts for material, size, set. ______________
14. Touch up shop-primed elements as required. ______________
15. All components adjusted, tested, and lubricated. ______________
16. Surfaces clean. ______________
17. All components free from discoloration, scratches, dents, and defects. _____________
18. Written warranty received. ______________
19. Certificate from inspecting authorities received. ______________
20. Written maintenance provision received. ______________

General Notes:____

Inspected by:______________________Date:__________________

Figure 1.23 Hydraulic Elevators Checklist.

Contractor's Elevator Pre-Inspection Checklist

- ❑ 1. One smoke detector required in each elevator lobby and elevator machine room. Wiring from the detectors to run to the machine room to the elevator controller. Smoke detectors shall not be self re-setting. Primary and alternate zones for detectors are required to provide the code-required elevator alternate landing feature.
- ❑ 2. Metal pit ladder is to extend from pit floor upwards, not less than 42" above the bottom landing floor level.
- ❑ 3. Pit light and switch with convenience outlet; switch shall be accessible from pit access floor. The pit convenience outlet shall have Ground Fault Protection (GFIC).
- ❑ 4. Machine room vented, if necessary, to maintain temperature between 60 and 100 degrees F. A type ABC fire extinguisher is required.
- ❑ 5. Fused mainline disconnect switch, lockable in the "off" position, in the machine room with feeder wires to elevator controller, all piped according with NEC and grounded. Disconnect switch must be in sight of elevator machine and shall be of type that cannot be engaged with the door open.
- ❑ 6. One (1) 120 volt, 20 amp, single phase power supply, fused and lockable in the "off" position, in the machine room and run to the elevator controller for car light supply for each elevator.
- ❑ 7. Provide an ADA-compliant telephone or intercom in the elevator cab that is hooked up to a 24-hour maintained location.
- ❑ 8. Only elevator equipment is allowed in the elevator machine room. If a sprinkler head is located in the machine room, it must have a shut-off valve located outside the machine room and marked for elevator machine room. When hoistways and/or machine room sprinklers are provided, then an automatic disconnect for elevator power (shunt trip) must be provided.
- ❑ 9. Machine room door - B-Labeled - shall be self-closing and self-locking.
- ❑ 10. Elevator hoistway shall be two (2) hour rated. Machine room(s) shall be rated for two (2) hour fire rating. There are exceptions to the rule, but it varies between areas.
- ❑ 11. Pit shall be so designed as to prevent the entry of ground water and remain dry. A drain or sump pump is required and the sump pump recess shall have a metal cover, flush with the floor.
- ❑ 12. All machine rooms must have permanent lighting (10-foot candles at floor).
- ❑ 13. Hoistway walls shall be substantially flush on hoistway sides. Any offsets over 2" shall be provided with a beveled angle of not less than 75 degrees.
- ❑ 14. Pipes or ducts conveying gases, vapors, or liquids which are not used in connection with the operation of the elevators are not permitted in hoistways or machine rooms.
- ❑ 15. Spaces containing machines, control equipment, sheaves, and other machinery shall be enclosed with fire-resistive enclosures. Enclosures and access doors there shall have a fire-resistant rating at least equal to that required for the hoistway enclosure.
- ❑ 16. Grout space between floor and sill edge.
- ❑ 17. Patch any holes in the elevator hoistway wall(s).
- ❑ 18. Hoist beam located per approved elevator layout drawing.

Figure 1.24 Contractor's Elevator Pre-Inspection Checklist.

Inspection Checklist

Interior Glass and Glazing **Project No:**____________

1. Approved submittals: shop drawings, samples, product data are on site. __________
2. UL labels are required. __________
3. Materials are properly stored on site and protected. __________
4. All material furnished and of approved types, thickness, and sizes. __________
5. All accessory items and glazing materials furnished. __________
6. Environmental: climatic and temperature conditions are suitable for installation. __________
7. Substrates and surfaces to be glazed are clean and primed as required. __________
8. Rabbets filled without voids. __________
9. Glazing gaskets complete bond. __________
10. Glass set tight and straight with proper bite and adequate clearance. __________
11. Glazing materials trimmed and cleaned from glass, stops, frames. __________
12. No rattling, no looseness. __________
13. Glass clean, no face imperfections, no glass edge damage. __________

General Notes:____

Inspected by:____________________Date:______________

Figure 1.25 Interior Glass and Glazing Checklist.

Inspection Checklist

Landscaping – Plantings **Project No.____________**

1. Observe that the layout of landscape construction such as walls, fences, paving, paths, benches are as required. ____________

2. Observe that the contractor verifies grades and elevations. Record field conditions as required. ____________

3. Plantings areas are in horticultural condition required. ____________

4. Observe that required drainage conditions are provided. ____________

5. Observe layout of major plant materials and adjustment to field conditions is provided as required. ____________

6. Plant materials are approved and inspected before installation. ____________

7. Staking, pruning, spraying, etc. are as required. ____________

8. Plants are protected from acid backsplash used to clean masonry or concrete. ____________

9. Do not allow diluted masonry cleaners to be dumped near root zones of trees or plantings. ____________

10. To avoid scorching of foliage, hot tar boilers are not allowed near trees or vegetation. ____________

11. Maintenance is provided as required during construction. ____________

12. Arrangements with hand-over and start of permanent maintenance are coordinated. ____________

General Notes:____

Inspected by:________________________Date:________________

Figure 1.26 Landscaping - Plantings Checklist.

Inspection Checklist

Landscaping - Finish Grading **Project No.**____________

1. Subgrade is scarified as required to receive finish grading. ____________

2. Stockpiled topsoil is distributed, prepared, and of depths required. ____________

3. Imported, borrowed, or selected off-site soils used as required. Get certificates, weight tags, etc. if required. Source is as required. ____________

4. Topsoil mixture and preparation is as required. ____________

5. Observe that the installation of sleeves, raceways, boxes, and piping required for irrigation and drainage. Electrical is provided as required. ____________

6. Verify that contractor has coordinated the installation with site improvements, paving, walks, etc. ____________

7. Areas to receive plantings are not excessively compacted by traffic, storage, or equipment. ____________

8. Scheduling of phases of landscaping work is in accordance with overall construction. ____________

9. Work is scheduled to avoid out-of-phase sequences that could cause damage or require re-work. ____________

General Notes:____

Inspected by:____________________Date:____________

Figure 1.27 Landscaping - Finish Grading Checklist.

Inspection Checklist

Landscaping - Sitework and Existing Vegetation **Project No.________**

1. Existing trees and other vegetation to remain are protected as required. ____________
2. Barricades and fencing are provided, if required. ____________
3. Traffic, parking, and storage of materials or debris is not allowed within drip line of existing trees to remain. ____________
4. Existing trees and other vegetation are maintained by regular feeding and watering during construction. ____________
5. Pruning of branches is performed only by qualified persons. ____________
6. Observe excavation adjacent to existing trees. Unless otherwise required, do not allow exposing of root system; keep equipment beyond drip line. ____________
7. Ponding around base of trees does not occur. ____________
8. Depth of cuts and other conditions required are met for the retainage and/or re-use of on-site topsoil. ____________
9. Observe required stockpiling and that topsoil is not mixed with deleterious material. ____________
10. Subgrade is carried sufficiently below finish grade to provide depth of topsoil required. ____________
11. Do not allow debris in subgrade that is detrimental to planting. ____________
12. Observe that elevations of manholes, catch basins, valves, boxes, etc. are coordinated to finish grades required on landscaping plans. ____________
13. Verify that contractor has coordinated work with related trades. ____________
14. Backfill against foundation walls that receive planting is clean and free of rocks, concrete, and other debris. ____________
15. Paints, solvents, oils, and debris are not placed in areas to receive planting. ____________
16. Do not allow construction equipment to operate and compact areas to receive plantings. ____________

General Notes:____

Inspected by:__________________________Date:__________________

Figure 1.28 Landscaping - Sitework and Existing Vegetation Checklist.

Inspection Checklist

Lockers and Benches **Project No.** __________

1. Approved submittals, shop drawings, samples, and product data are on site. __________
2. Materials are properly stored on site and protected from damage. __________
3. All materials furnished are approved types, sizes, gauge, and colors. __________
4. All accessory items furnished are approved types. __________
5. All parts inspected after unpacking for damage, dents, scrapes, etc. __________
6. Observe coordination and scheduling with other trades including rough carpentry and drywall, CMU walls. __________
7. Lockers and benches have required structural attachment at walls and floor. __________
8. Metal bases and sloping tops installed as required. __________
9. Metal filler panels provided as required. __________
10. Flush, hairline joints provided against all adjacent surfaces. __________
11. Lockers and benches plumb, level, rigid, and flush. __________
12. Doors, latches, integral locking devices adjusted for ease of operation. __________
13. Touch-up shop painted areas as directed. __________
14. Lockers and benches cleaned and labels removed__________
15. Keys identified and turned over to owner's representative. __________

General Notes:____

Inspected by:________________________Date:________________

Figure 1.29 Lockers and Benches Checklist.

Inspection Checklist

Louvers **Project No.** ____________

1. Approved submittals, shop drawings, samples, and product data on site. ____________
2. Materials properly stored and protected. ____________
3. All materials furnished and of approved types, sizes, shapes, and colors. ____________
4. All accessory items furnished and approved. ____________
5. Observe coordination and scheduling with other trades including misc. metal, structural steel, masonry, sealants, and mechanical. ____________
6. Check anchoring devices and inserts for material, size, and set. ____________
7. Bolts, brackets, sleeves embedded in concrete galvanized. ____________
8. Louvers have suitable structural attachments. ____________
9. Welders, if required, are certified. ____________
10. Welds, if required, are continuous along entire line of contact. ____________
11. Exposed welds, if required, are flush and ground smooth. ____________
12. Threaded connections are tight, all threads concealed. ____________
13. Exposed bolts and screw heads are flat and countersunk. ____________
14. Exposed joints are not conspicuous and are close fitting. ____________
15. Louver units plumb, level, and in proper alignment. ____________
16. Shop painted prime and finished coats touched up as required. ____________
17. Louver free from scratches, waves, dents, buckles, and tool marks. ____________
18. Protective coverings used to guard work from construction abuse. ____________

General Notes:____

Inspected by:______________________ Date:________________

Figure 1.30 Louvers Checklist.

Inspection Checklist

Masonry **Project No.__________**

1. Approved shop drawings on site. __________
2. Approved samples on site or evidenced. __________
3. Sample panel, if required, constructed and approved. __________
4. Materials stored off ground and covered. __________
5. Correct type/color mortar. __________
6. Concrete masonry units are not wet. __________
7. Reinforcement: type, size on site, spacing specified. __________
8. Excessive bending of rebar not allowed. __________
9. Pipes, sleeves, boxes located. __________
10. No shovel measures for job-mixed grout. __________
11. Climatic and temperature controls are acceptable. __________
12. Adequate lighting available for good workmanship. __________
13. Joint size, type, tooling method as required. __________
14. Bonding is as required. __________
15. Observe full head and bed joints. __________
16. Joints tooled to provide dense surface. __________
17. Bond beams in place, properly reinforced and grouted. __________
18. Wythes or cavities kept free of excess mortar. __________
19. Check anchors & ties for materials, sizes, etc. __________
20. Bucks and anchors - secured, plumb, and level. __________
21. Provisions for flashings, cut-outs, & later items. __________
22. Provisions for parging, if required. __________
23. Expansion and control joints are located. __________
24. Structural members have suitable attachments. __________
25. Debris is removed periodically, not piled. __________
26. Protect work from freezing for at least 48 hours. __________
27. Clean off splatters. __________
28. Observe bond beam filling. __________
29. Hollow metal frames fully grouted, if required. __________
30. Backfilling after proper securing & support. __________
31. Proper support of high walls. __________

General Notes:____

Inspected by:____________________Date:______________

Figure 1.31 Masonry Checklist.

Inspection Checklist

Mechanical - General Items **Project No.**____________

1. All materials (equipment and items) delivered to site are inspected for damage in transit. ____________

2. All materials are identifiable by tags, markings, stamps, etc. showing size, material, gauge, weight, grade, capacity. ____________

3. Material are new, unless otherwise specified. ____________

4. Materials are adequately stored and protected. ____________

5. Installation is coordinated with other trades to avoid interference, damage. ____________

6. Approved shop drawings are on site before installation. ____________

7. Approved samples are on site or evidenced before installation. ____________

8. Approved color and material schedule is on site prior to installation. ____________

9. Existing and adjacent work connections and tie-ins are suitably performed and in location as required. ____________

10. Surfaces to receive materials are acceptable. ____________

11. Climatic and temperature conditions are suitable and as required before, during, and after installation. ____________

12. Adequate protection is provided for adjacent surfaces before, during, and after installation. ____________

13. Local public inspections and design engineer inspections have been performed as required. Reports have been issued/received. ____________

14. Work installed is cleaned and protected. ____________

15. Warranties and guarantees have been received. ____________

16. Spare parts, attic stock, as required, received. ____________

17. Operating & Maintenance manuals have been submitted. ____________

General Notes:____________

Inspected by:____________ Date:____________

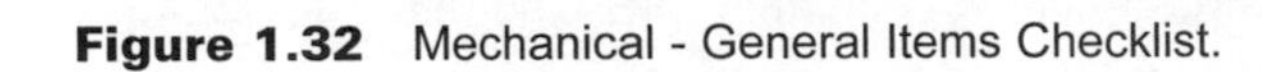

Figure 1.32 Mechanical - General Items Checklist.

Inspection Checklist

Mechanical - Power or Heat **Project No.____________**

Boilers, Equipment, and Distribution

1. Boilers conform with or are identified with ASME code. ____________

2. Bases or refractory bases are provided as required. ____________

3. Expansion joints are provided and guided. Check requirement for expansion joints in floor around boiler or generator. ____________

4. Gas burners - approved standard, position of pilot flame and sensing element, regulators, and controls provided. Regulator installed in vertical position and gas vents piped to exterior. ____________

5. Forced draft fans - check fans for these features: anchorage; alignment and rotation; accessibility for lubrication; damper operation as required; insulation application; safety control interlocks and airflow switches. ____________

6. Oil storage tank - approved standard tank capacity and calibration, required openings, proper anchorage, minimum cover and/or clearance, containment or double wall provided as required. ____________

7. Piping is of material, size, weight, and type required. Lines are reamed, openings are protected, and fitting connections are as approved. Swing joints, if required. ____________

8. Piping is connected with fittings required; tapered threads are used. Joints are welded with approved welding procedures and qualified personnel as required; piping is in grade and alignment without strain on fittings. ____________

9. Valves are installed of type and position required; all fittings such as strainers, checks, gauges, air reliefs, drips, traps are provided. ____________

10. Piping is painted and/or identified for flow and type as required. ____________

11. Shut-offs for fuel and water are provided. ____________

12. Valves are provided to shut down sections of system, if required. ____________

13. Valves are labeled, if required. ____________

14. Safety and relief valves are provided and set to required PSIG. Discharges are piped to drains. ____________

15. Safety operating controls are provided as required. ____________

16. Combustion air system is provided as required. ____________

17. Breaching and flues are of proper material, construction, and type and are installed as required. ____________

18. Expansion tanks are located, mounted, and anchored as required and are provided with accessories and drain. ____________

19. Valves and fittings are insulated as required. ____________

Figure 1.33 Mechanical - Power or Heat Checklist. *(continued on next page)*

Terminal Units

20. Heating and ventilating units - anchored and provided with vibration isolators as required; access doors are provided and fit tight; flexible connectors are provided as required; controls are provided as required; location and layout is coordinated. ____________

21. Unit heaters - noise level is within approved range; all clearances and location are as approved, adequate air distribution is provided; controls as required are provided. ____________

22. Baseboard units - location, type, size, mounting and controls are provided as required. Covers, access doors, dampers, end plates are provided to extent required. ____________

Cleaning, Testing, Balancing

23. Verify system is completely clean and flushed of all dirt and debris. ____________

24. Operate system in presence of inspectors or engineers as required. ____________

25. System is completely balanced; balancing report is approved. ____________

26. Location of all balancing devices and/or valve chart has been prepared and transmitted to the engineer. ____________

27. Engineer of record has accepted system and submitted report. ____________

General Notes:____

____

____

____

____

____

____

Inspected by:______________________________Date:__________________

Figure 1.33 *(continued from previous page)* Mechanical - Power or Heat Checklist.

Inspection Checklist

Mechanical Refrigeration **Project No.________**

1. All materials are as approved - nameplates, identification, and characteristics are provided and are not covered by insulation or painted out. ____________

2. All rotating parts, belts, etc. have guards or provide protection. ____________

3. Vibration isolators and flexible connections are provided as approved and required. ____________

4. Fire separation from fuel-fired equipment is provided if required. ____________

5. Freeze protection devices and materials provided if required. ____________

6. Tube removal, cleaning space, or provisions are adequate. ____________

7. Clearances to all equipment electric panels are adequate. ____________

PIPING

8. Types, size, weight, material, and fittings are as approved and required. ____________

9. Installation method of piping is as required. ____________

10. Pipes are square cut & reamed; soldering is as required; internal valve parts are protected against heat or removed; joints are thoroughly cleaned and fluxed and excess flux and acid are removed. Solder used is of approved type. ____________

11. Flexible connections are provided as required. ____________

12. Unions and flanges installed for maintenance are as required. ____________

13. Lines are properly pitched as required. ____________

14. Air vents are installed at high points & drains installed at low points as required in water lines. ____________

15. Proper type of valves are provided, i.e. gate or globe. ____________

16. Balancing cocks are installed as required. Pressure gauges, thermal elements, etc. are provided. ____________

17. Hangers are provided - anchored, installed, and of type, number, and spacing required. See that dissimilar-metal isolation is provided. If hangers are to be installed over insulation, see that high-density insulation inserts and metal shields are provided as required. ____________

18. System is checked for leaks. ____________

19. Insulation is provided and installed as required. Vapor barriers, adhesives, and sealants are non-combustible. ____________

20. Requirements for insulating flanges, fittings, and valves are met. ____________

21. Piping is properly sealed and flashed as required when penetrating building elements. ____________

22. Cooling coil condensate drawings with trap seats are provided if required. ____________

Figure 1.34 Mechanical Refrigeration Checklist. *(continued on next page)*

EQUIPMENT

23. Air-cooled condenser - air flow is not obstructed and wind deflectors are provided if required. ____________

24. Evaporator condenser - check for spare coverage, quiet float valve, and water level. ____________

25. Reciprocating compressor - check for shaft alignment on direct drive. Also suction and discharge pressures, installation of required gauges, motor amperage under maximum load, cylinder head overheating. ____________

26. Screw compressor – check for alignment of unit and drive, noise and vibration, safety controls, shaft seats as required. ____________

27. Receiver location is out of direct sun if installed outside building. ____________

28. Relief valve on receiver is of size required and discharges to atmosphere. ____________

29. Receiver drain, purge valve, liquid level indication, and shut-off valves are provided and/or as required and piped to exterior as required.__________

30. Cooling tower - location and provision for mounting are as approved and required. Overflow and drain piping are provided. ____________

31. Mechanical draft cooling tower - unobstructed air intake provided, fan rotation and speed, belt tension as required, freeze protection for piping, traps, basin and pump, vibration switches, weather protection is provided for motor if required. Catwalk protection is installed. ____________

32. Pumps are supported properly, free of excess vibration. Piping around is adequately supported. All gauges and motors are provided. ____________

33. All insulating materials are provided and installed as approved. ____________

34. Observe procedures for testing of systems. ____________

General Notes:____

____

____

____

____

____

____

____

Inspected by:________________________Date:____________________

Figure 1.34 *(continued from previous page)* Mechanical Refrigeration Checklist.

Inspection Checklist

Membrane Roofing **Project No.____________**

1. All shop drawings, product data, samples as required and approved are on site. ___________

2. Attend or conduct preconstruction meeting. Review construction sequence, any deviations from approved submittals, field problems, warranty issues. ___________

3. Before roofing contractor commences work, observe the following:

   a. Surfaces are clean and free of debris. ___________

   b. Excess mortar or concrete is removed. ___________

   c. All holes, joints, and cracks are pointed. ___________

   d. All rough or high spots are ground smooth on concrete deck. ___________

   e. Wood nailers or other attachment conditions are adequate. ___________

   f. Surfaces are dry as required by manufacturer. ___________

   g. Concrete deck tested for dampness if necessary. ___________

   h. Slope is as required. ___________

   i. Pipes, conduits, and other roof penetrations are in place and ready to receive flashings. ___________

   j. All sheet metal and roof accessories are in place or on hand to be installed as the roof installation commences. ___________

4. Materials of types required are provided. Materials are identifiable and comply with ASTM standards. ___________

5. If rolled roofing, stand rolls on end and keep free of moisture. ___________

6. Nails and fasteners are of length, shank, head, and coating required. ___________

7. Surface to receive roofing is primed, if required. ___________

8. Observe lap, nailing, and quantity of adhesive applied. ___________

9. See that membrane is laid so that it is free of air pockets, wrinkles, and buckles. If present, rolling may be required. ___________

10. All surfaces to be kept free of moisture. Under no condition allow exposure of insulation or incomplete install membrane to remain overnight without protection. Protect stored material from moisture. ___________

11. Observe installation of cant strips, vertical surfaces, reglets, and penetrations. ___________

12. Observe sealing of roof membrane envelopes where the use of envelope is permitted. ___________

13. All concrete walls to receive roofing are primed. ___________

14. Observe welds; do not allow any skips or unwelded joints. ___________

15. Avoid plugging of drains and weeps and do not damage adjacent surfaces. ___________

Figure 1.35 Membrane Roofing Checklist. *(continued on next page)*

16. Roofing membrane is fully set into clamping ring. Lead collar flashing is installed and stripped in, if required. ___________

17. Roofing is protected against damage by other trades. ___________

18. Observe cut samples, if required. Observe that patching of cut samples is properly performed where samples are cut. ___________

19. Clean-up provided after installation; drains cleared. Debris removed from site. ___________

General Notes:___________

Inspected by:___________Date:___________

Figure 1.35 *(continued from previous page)* Membrane Roofing Checklist.

Inspection Checklist

Metal Toilet Partitions **Project No.**______________

1. Approved submittals, shop drawings, product data on site. ______________
2. Materials are properly stored on site and protected. ______________
3. All materials furnished and of approved types, sizes, gauges. ______________
4. All accessory items and hardware furnished and of approved types. ______________
5. Observe coordination with other trades including masonry, drywall, ceiling work, tile, plumbing. ______________
6. Partitions have required structural attachments at wall, floor, and/or overhead bracing as required. ______________
7. Partitions and doors are rigid, straight, plumb, and level. ______________
8. Proper clearances, as required. ______________
9. Final adjustments performed - pilaster, leveling devices, door hardware. ______________
10. Hardware lubricated. ______________
11. Minor scratches and small imperfections touched up. ______________
12. All exposed surfaces cleaned and labels removed. ______________

General Notes:____

Inspected by:________________________Date:______________

Figure 1.36 Metal Toilet Partitions Checklist.

Inspection Checklist

Metal Deck **Project No.**_____________

1. Approved shop drawings are on site. _______________
2. Material approved, type, gauge, finish, shape, size. _______________
3. Verify that contractor has reviewed approved decking layout submissions. ___________
4. Material is properly stored and protected. _______________
5. All accessory items are furnished as approved. _______________
6. Welding inspection by testing lab is required. _______________
7. Welders are certified, if required. _______________
8. Sequence of fastening is performed. _______________
9. Closures at edges are provided as required. _______________
10. Decking in contact with beams; proper tack welds. _______________
11. Observe tabs and hangers for equipment and ceilings. _______________
12. Observe coordination with sheet metal, roofing, insulation, MEP work. _______________
13. Decking is continuous over supports when required. _______________
14. Welded connections and spacings are as required. _______________
15. Observe panel-to-panel seams for tack welds. _______________
16. Check seam welding for burn-outs. _______________
17. Reinforcement at columns and penetrations, as required. _______________
18. Reinforcement at major concentrated loads as required. _______________
19. Type, spacing, alignment of seam connections as required. _______________
20. No rough edges to damage wire pulled through cells. _______________
21. Butt ends taped to keep concrete fill out. _______________
22. No concentrated loads on decks during construction. _______________
23. Verify if shoring is required for concrete topping. _______________
24. Deck is free of loose dirt and debris before topping. _______________
25. On roof decking, de-slag welds and paint with zinc primer.____________________
26. Verify if U.L. labels are required. _______________
27. Roof ventilation provisions are met. _______________
28. All deck openings properly reinforced. _______________
29. Touch-up paint on primed deck/zinc on galvanized. _______________
30. Record all damaged panels. _______________

General Notes:____

Inspected by:__________________________Date:__________________

Figure 1.37 Metal Deck Checklist.

Inspection Checklist

Metal Framing - Gypsum Drywall **Project No.____________**

Metal Framing

1. All submittals, samples, shop drawings are approved and on site. ____________
2. Material is stored in a dry location. ____________
3. Material galvanized as required. ____________
4. Studs are doubled up at jambs, unless otherwise required. ____________
5. Structural and/or heavy gauge studs as required. ____________
6. Studs allow for movement, slab deflection. ____________
7. Studs securely anchored to walls, columns, and floors. ____________
8. Soundproofing provided at floor and walls as required. ____________
9. Observe location, layout, plumbness. ____________
10. Channel stiffeners are provided as required. ____________
11. Special fastening and connections are observed. ____________
12. Anchor blocking, plates, other equipment provided. ____________
13. Cut studs for openings are properly framed. ____________
14. Observe size, gauge of runner, and furring channels. ____________
15. Hangers are saddle tied, bolted, or clipped as required. ____________
16. Tie wire for channels to runners properly tied. ____________
17. Elevation and layout of furring is understood. ____________
18. Observe that surfaces are plumb and level. ____________
19. Observe that long single lengths are used. ____________
20. Control joints are installed per contract requirements. ____________
21. Requirements for adjoining surfaces of different materials are accommodated. ____________
22. Seating provided for sound or thermal insulation. ____________
23. Spacing and construction are as specified. ____________
24. Observe location of all blocking, bracing, and nailers. ____________
25. Type, thickness, length, and edges are as required. ____________
26. Type fastener, length, and spacing as required. ____________
27. Installation complies with manufacturer's requirements. ____________
28. Special type suited for damp locations if required. ____________
29. Special lengths are provided as required. ____________
30. Verify if horizontal or vertical application is required. ____________
31. Wall board is installed with staggered application. ____________
32. Internal and external metal/plastic corners as required. ____________

Figure 1.38 Metal Framing - Gypsum Drywall Checklist. *(continued on next page)*

33. Number of coats of compound required is provided. ______________
    Level 1 - All joints and interior angles have tape embedded in one layer of joint compound. ______________
    Level 2 - All joints and interior angles taped & receive two coats of taping compound. ______________
    Level 3 - All joints and interior angles taped & receive three coats of taping compound. ______________
    Level 4 - Apply Level 3 plus skim coat of joint compound over entire surface. ______________
34. Sanding between coats is performed. ______________
35. Feathering is out 12" to 16". ______________
36. Provide air circulation with adequate dry heat. ______________
37. If fire rated, recesses over 16" are boxed in. ______________
38. Penetrations tight and sealed as required by code. ______________
39. Verify contractor has coordinated cut-outs and outlet boxes correctly to avoid patching. ______________
40. Wall board is held up from floor 3/8" minimum. ______________
41. Vertical joints are aligned with door jambs. ______________
42. Damaged sheets are not used and are removed. ______________
43. Observe minimum piecing or joining. ______________
44. Non-metallic cable, plastic, or copper pipe is not damaged. ______________
45. Check for bubbles and dimples. ______________
46. Curing time is adequate for subsequent finishes. ______________

General Notes:____

Inspected by:______________________ Date:______________

Figure 1.38 *(continued from previous page)* Metal Framing - Gypsum Drywall Checklist.

Inspection Checklist

Metal Stairs **Project No.**______________

1. Approved submittals, shop drawings, samples, and product data on site. ____________
2. Stamped calculations in reference to structural properties on site. ____________
3. Materials are properly stored on site and protected. ____________
4. All materials furnished are approved types, shapes, gauges, metal treatments, and finishes. ____________
5. All accessory items furnished are of approved types. ____________
6. Observe coordination and scheduling of the work of related trades including misc. metal, and structural steel framing. ____________
7. Welding inspection by testing laboratory, if required. ____________
8. Welders are certified. ____________
9. Welds continuous along entire length of contact. ____________
10. Exposed welds are flush and ground smooth. ____________
11. Exposed joints not conspicuous, close fitting. ____________
12. Exposed bolts and screws cut off flush with nuts or other adjacent surfaces. ____________
13. Threaded connections tight - all threads concealed. ____________
14. Exposed bolts and screw heads flat and countersunk. ____________
15. Shop painted coats touched up as required. ____________
16. Surfaces clean of stains, grease marks removed as required. ____________
17. Metal stairs free from scratches, waves, dents, buckles, tool marks, and rattles. ____________

General Notes:____

____

____

____

____

____

____

____

Inspected by:______________________ Date:______________

Figure 1.39 Metal Stairs Checklist.

Inspection Checklist

Overhead Coiling Doors **Project No.**______________

1. Approved submittals, shop drawings, samples, product data, and maintenance instructions, as required, are on site. ______________
2. Materials and components are properly stored and protected on site. ______________
3. UL label, if required, is affixed to product. ______________
4. All material furnished and approved type, shapes, gauges, and sizes. ______________
5. All accessory items furnished and of approved types. ______________
6. Observe coordination and scheduling with: concrete, masonry, rough carpentry, miscellaneous metals, finish hardware, electrical. ______________
7. Check anchoring devices and inserts for material, size, and set. ______________
8. Jambs, head mold strips, and anchors secured, plumb, level. ______________
9. Structural members have suitable attachments as required. ______________
10. Doors adjusted, tested, lubricated. ______________
11. Watertight fit. ______________
12. Electrical operation (if required), correct amps/voltage compatible with panel requirements, operator functions properly. ______________
13. Touch-up paint as required. ______________
14. Doors cleaned. ______________
15. Written warranty received, reviewed, per specifications. ______________

General Notes:____

Inspected by:____________________________ Date:___________________

Figure 1.40 Overhead Coiling Doors Checklist.

Inspection Checklist

Painting **Project No.____________**

1. Color scheme is complete and understood. ______________
2. Approved product data, shop drawings, samples are on site. "Paint-outs" are matched against samples. ______________
3. Prior understanding is made on stopping points for change of color or finish. ______________
4. All materials are new and products of same manufacture, if required. ______________
5. Containers are adequately identified. Reject containers with seals broken. ____________
6. Surfaces to receive paint are dry. Moisture meter tests on plaster, concrete, or masonry surfaces are made, if required. ______________
7. Damp, not wet, surfaces are allowed for water-thinned paints. ______________
8. Surfaces to receive paint are sanded, holes puttied or filled. ______________
9. Pitch pocket, knots, and shakes are shellacked or treated and otherwise cleaned of deleterious substances. ______________
10. Metal surfaces are treated, primed or otherwise cleaned as required. ______________
11. Areas are suitably cleaned and free of conditions affecting drying. ______________
12. Dust control is maintained. ______________
13. Temperature conditions for type of paint are provided. ______________
14. If heating required, it has been sufficiently provided in advance in order to have surfaces up to temperature and avoid condensation. ______________
15. Adequate lighting is provided for proper working conditions. ______________
16. Protection of adjacent areas, surfaces, and items is provided. ______________
17. Hardware, trim, and fixtures are removed for painting or otherwise protected. ______________
18. Observe, occasionally, the mixing and thinning of paints. Thinning should be controlled and the need demonstrated. ______________
19. Required number of coats is provided. Tinting of undercoats is provided if required. Opacity is being achieved. ______________
20. Required texture and method of application - spray, brush, roller - is understood and being employed. ______________
21. Lumps or bumps do not appear on applied coats. ______________
22. Workmanship is adequate. Do not allow runs, drips, laps, brush marks, "lace curtains" or "holidays" variations in color, texture, finish. ______________
23. Doors receive coats on both faces at essentially the same time. ______________
24. Observe that all edges and specifically top and bottom edges are painted. ___________

Figure 1.41 Painting Checklist. *(continued on next page)*

25. Curing time required between coats is provided. ______________
26. Sealers, fillers, and stains are applied and treated as required. ______________
27. Putty is not applied until after staining or priming and matches wood. ______________
28. Hard-to-get-places are painted - bottom of shelves, back of trim in corners. ______________
29. Correction of unsuitable work is made promptly. ______________
30. Proper types and quantities of "attic stock" provided to Owner. ______________

General Notes:____

Inspected by:______________________Date:______________

Figure 1.41 *(continued from previous page)* Painting Checklist.

Inspection Checklist

Pavement and Walks **Project No.____________**

1. Observe that contractor has installed sub-grade to proper elevation. ____________
2. Sub-grade is dense and properly compacted. ____________
3. All drains, utilities, and other underground construction are properly coordinated and in place. ____________
4. Trench backfilling is performed as required. ____________
5. Control testing of sub-grade and sub-grade materials is being performed and recorded if required. ____________
6. Sub-grade and base courses are of type, thickness, and material specified. ____________
7. Sources of materials have been sampled and approved. ____________
8. Location of all manholes, outlets, and other surface features is known and properly coordinated and placed. ____________

Concrete Paving:

1. Contractor has provided approval of mix design, batch plant. ____________
2. If admixtures are furnished, provide engineer approval. ____________
3. Base course is maintained in a firm, moist condition. ____________
4. Observe that all forms, headers, outlets, boxes, equipment are in place prior to pour. ____________
5. Observe that all embedded items, sleeves, dowels, and reinforcements are installed as required and in their proper position/elevation. ____________
6. Joint methods and materials are provided and observed. ____________
7. Grade, slope, pitch, and thickness control are provided. ____________
8. Concrete is deposited, rodded, or vibrated to suit conditions. ____________
9. Observe that reinforcement is maintained at the required elevation. ____________
10. Time intervals between pours allow for continuous working, no cold joints. ____________
11. Control joints, construction joints, and expansion joints provided as required. ____________
12. Finishing treatment and texture is as required. ____________
13. Curing provisions are as required and the work is properly protected. ____________
14. Overtroweling is to be avoided. ____________
15. Jointing of new concrete to old is performed properly. ____________
16. Forms are not to be removed until minimum time after placement has been achieved. ____________
17. Sawed joints to be made of proper width, depth. ____________

Figure 1.42 Pavement and Walks Checklist. *(continued on next page)*

18. Joints to be cleaned and cured as required. ____________
19. Joints to be properly sealed. ____________
20. Concrete to be protected during backfill operations. ____________
21. Test for proper drainage before acceptance. ____________

General Notes:____

Inspected by:______________________________Date:__________________

Figure 1.42 *(continued from previous page)* Pavement and Walks Checklist.

Inspection Checklist

Resilient Flooring **Project No.** ______________

1. Samples, product data are approved and on site. ______________
2. Type, pattern, and color of material is as approved. ______________
3. Primer, adhesive, or cement is as required. ______________
4. Substrate is inspected and acceptable for installation:
   If MDF on wood subfloor:
   a. Underlayment properly fastened to subfloor. ______________
   b. All joints securely fastened. ______________
   c. Sufficient gap at joints to allow for expansion. ______________
   d. Subfloor free from squeaks, soft spots. ______________
   If concrete:
   a. Concrete properly cured. ______________
   b. Moisture test is in acceptable range. ______________
   c. Large chips/cracks properly filled with flash patch material. ______________
   d. Any humps ground down and depressions filled to level. ______________
5. Containers labeled. Sealed and unopened. ______________
6. Boxes inspected for damage in transit. ______________
7. Floor material stored at proper temperature. ______________
8. Pre-formed corners and end stops are proved. ______________
9. Division strips as required are on site. ______________
10. Work is being installed in proper sequence. ______________
11. All areas are cleaned before installation. ______________
12. On slabs, surfaces are primed if required. ______________
13. Area temperature maintained during installation. ______________
14. Direction of tile is as specified. ______________
15. Cement applied at proper rate and proper tackiness. ______________
16. Joints and seams are tight and level. ______________
17. Minimum length of cuts is observed. ______________
18. Observe provisions made for thresholds and joinings. ______________
19. Observe level joining at any floor escutcheons or plates. ______________
20. Excess adhesive promptly removed. ______________
21. Scuffed, broken, or discolored tile is replaced. ______________
22. Temporary protective cover is provided. ______________
23. Surface is cleaned and sealed/polished as required. ______________
24. Warranties/guarantees provided as required. ______________
25. Attic stock, as required, is supplied. ______________

General Notes: ____

____

____

____

Inspected by: ______________________ Date: ______________

Figure 1.43 Resilient Flooring Checklist.

Inspection Checklist

Roof Insulation **Project No.**____________

1. Approved shop drawings, product data, samples on site. __________
2. Materials are of type required for surfaces, treatment, ratings, sizes, thickness, etc. __________
3. Materials are stored to prevent moisture infiltration and are installed dry. __________

If Rigid Insulation:

4. Wood nailers/stops are provided at perimeter as required. __________
5. Vapor barrier, if required, is provided. Observe installation, nailing requirements. Check that vapor barrier seals insulation at gravel stops, walls, openings. __________
6. Method of installing over decking, under topping, vapor barrier is as required. __________
7. Fasteners, when specified, are of proper type, size and spacing. __________
8. Observe fastener penetration through decking if required. __________
9. Joints are staggered, except when joints are to be taped. When two layers are installed, vertical joints should be offset. Do not allow joints over flute openings in steel deck. __________
10. Insulation is installed in conjunction with roofing membrane when required. __________
11. Water cut-offs, if required, are installed at end of each day's work. __________
12. Insure insulation is covered by roofing each day. __________
13. Fire-resistant adhesive is provided where required. __________
14. Insulation remains dry until covered by roofing. __________
15. Insulation at roof drains should permit proper drainage. May require routing or back cutting. __________

General Notes:____

Inspected by:____________________ Date:____________

Figure 1.44 Roof Insulation Checklist.

Inspection Checklist

Rough Carpentry: **Project No.__________**

1. Delivered lumber is of proper species and grade and has treatment required. __________
2. Framing lumber is grade stamped or suitably identified. __________
3. Inspect for splits, checks, crook, warp, loose knots, decay, and pockets. __________
4. Lumber suitably stored off ground, stacked to prevent warp, protected to prevent increase in moisture content. __________
5. Grade stamp indicates moisture content as required. __________
6. Preservative treatment, if required, and affidavits enclosed. __________
7. Materials in contact with masonry or concrete, or near earth, are treated or of suitable species of lumber for those conditions. __________
8. Surfaces to be painted are treated with proper preservatives. __________
9. Framing is in alignment, plumb, level. __________
10. Nails, bolts, and connectors are as required and specified. __________
11. Spacing of fasteners per code or "best practices". __________
12. Allowance is made for expansion or contraction of lumber, concrete, or masonry. __________
13. Observe that bridging, blocking, and bracing are provided as required. __________
14. In-wall blocking provided for all equipment to be attached. __________
15. Plates are properly lapped and properly connected. __________
16. Connections to metal are as required. __________
17. Framing members are doubled where required. __________
18. Framing members are spaced as required; plumb, horizontal, parallel, and aligned. __________
19. Headers are of size required, have proper bearing, and are suitably connected. __________
20. Plywood/OSB sheathing applied as specified: grade, dimension, staggering, nailing, blocking, etc. __________
21. Clearances are provided, such as 2" on hot pipes and flues. __________
22. Furring and grounds are as required, properly aligned and plumb. __________
23. Sealing, especially for acoustical or waterproofing purposes, is provided where required. __________
24. Sheathing paper or air infiltration paper is provided, as required, without tears or other visible damage. __________
25. Agency or A/E inspections required before closing in.

General Notes:____

____

____

____

____

Inspected by:______________________________Date:____________________

Figure 1.45 Rough Carpentry Checklist.

Inspection Checklist

Steel Joists **Project No.**___________

1. Joists are of type and size per approved shop drawings. ___________
2. Joists are coated with types of paint and number of coats required. ___________
3. Verify that welds have been inspected for length and size. ___________
4. Nailer on top anchor and/or bottom chord is provided, if required. ___________
5. Joists are accurately spaced and have proper bearing and anchorage. ___________
6. Installation and connections are as required. ___________
7. Ceiling extensions are provided where required. ___________
8. Bridging and anchoring are installed as soon as joists are placed and before application of any loads. ___________
9. Ends of bridging lines terminating at walls or beams are anchored at plane of top and bottom chords as required. ___________
10. No cutting or drilling of web or chord members allowed. ___________
11. Do not allow excessive concentrated loads of heavy building materials or moving of any heavy equipment over joists. ___________
12. All rust, scale, slag, and splatter is removed and joist is clean before it is painted. ___________

General Notes:___________

Inspected by:___________ Date:___________

Figure 1.46 Steel Joists Checklist.

Inspection Checklist

Structural Steel **Project No.______________**

1. Approved shop drawings on site. ____________
2. Setting of foundation anchor bolts, size, location surveyed. ____________
3. Size and types of bolts/washers on site. ____________
4. Tension control (TC) bolts as required. ____________
5. Shop painting as required. ____________
6. Embedded items shop coated properly, if required. ____________
7. Delivered steel is new, undamaged, free of distortion. ____________
8. Proper tensile strength as required. ____________
9. Steel is suitably stored, blocked off ground, covered if prolonged storage. ____________
10. Column ends are milled and protected, if required. ____________
11. Observe setting of base plates. Full engagement of nut, no heating or bending of anchor bolts has occurred, no undue chipping of concrete. ____________
12. Temporary connections to hold steel in place are provided before final bolts or weld connections are made. ____________
13. Concrete is cleaned and free of dirt/debris and grouting is performed properly. ____________
14. Space between concrete and bottom of bearing plate usually must not exceed 1/24 bearing plate width. ____________
15. See that dry pack mortar is properly rammed. ____________
16. Camber is furnished as required. ____________
17. Beam members are set with natural camber up. ____________
18. Steel members are not field cut for passage of conduit, pipes unless approved by inspecting engineer. ____________
19. Type, size, and length of bolt and type of washer, and size of hole are all as required. ____________
20. All bolts, heads, and nuts inspected after installation. ____________
21. Burning of holes to correct misalignment not allowed. ____________
22. Verify whether paint is allowed on contact surfaces. Generally all deleterious materials, dirt, oil, loose scale or defects such as burrs or pits are not allowed. ____________
23. Slope of flanges (1:20), beveled washers as required. ____________
24. Hardened washers are provided as required. ____________

General Notes:____

____

____

____

Inspected by:________________________Date:________________

Figure 1.47 Structural Steel Checklist.

Inspection Checklist

Tile - Walls and Floors **Project No.____________**

1. Shop drawings, product data sheet, samples approved and on site. ____________

2. Containers are sealed with Grade Seals identifying grades of tile as required for glazed interior and ceramic mosaic tiles (quarry, glass mosaic). Cement body and marble tiles are not grade sealed. ____________

3. Tile color, sizes, patterns, shapes, and type are as approved. ____________

4. Trim shapes are appropriate for use as required, bullnose edges for thin set. ____________

5. Mastergrade certificates are delivered with tile shipment. ____________

General

6. Grout type and color as approved. ____________

7. Tile color is uniform, shadings within acceptable range. ____________

8. Layout is approved. Generally no cuts smaller that half tile size with cuts balanced and areas centered. ____________

9. Pattern of layout is as approved. ____________

10. Tile joints are straight and true. ____________

11. Tile surfaces are true to plane, level, and plumb. ____________

12. Tile corners are flush or level with adjacent tile. ____________

13. Tiles edges are on an even plane, smooth to touch. ____________

14. Tile cuts are smooth without jagged or flaked edges. ____________

15. Bonding is complete and sound. ____________

16. Finished tile is free from pits, chips, cracks, or scratches. ____________

17. Grout is uniform in color, texture, tooled uniformly on cushion edges tiles, flush to top of square edged tiles and smooth without voids. Is hard and durable. ____________

18. Finished tile surface is cleaned of setting and grouting materials. Do not permit use of Muriatic acid on glazed tile. ____________

19. Finished tile is protected from damage. ____________

20. Tile is laid with metal jigs or paper facing as required. ____________

21. Surface-applied accessories are fastened through tile by drilling tile or grout joints without inducing cracks. ____________

Figure 1.48 Tile - Walls and Floors Checklist. *(continued on next page)*

Wall Tile Thinset

22. Wall backing is properly secured and of specified type; water resistant if drywall; surface sealed as required, drywall penetration caulked with special sealants. ____________

23. Setting materials are as specified: organic adhesives on drywall; dry set or latex-portland cement on drywall, masonry, or cured mortar bed. ____________

24. Wall tile is set before floor tile. ____________

25. Built-in accessories are set and firmly anchored, aligned with grout joints as required. ____________

26.Tile is protected from movement, 48 hours after setting and 48 hours after grouting; no work on opposite side of wall. ____________

Floor Tile

27. Non-slip tile is provided for required areas. ____________

28. Marble thresholds are set to finish grade. ____________

29. Floor drains are set to finish grade at low point. ____________

30. Floor is uniformly sloped to drain as required. ____________

31. Expansion joints are located and of type required. ____________

32. Substrate is clean and dry without bumps and hollows for thinset application. ____________

33. Waterproof membrane is properly installed. ____________

34. Bond coat is per manufacturer's recommendations. ____________

35. Tile is tamped in to assure good bond; excess bond coat immediately removed before set-up. ____________

36. Tile is protected from traffic and dirt, 48 hours after setting and 48 hours after grouting. ____________

General Notes:____

____

____

____

____

____

____

____

____

Inspected by:______________________________ Date:________________

Figure 1.48 *(continued from previous page)* Tile - Walls and Floors Checklist.

Inspection Checklist

Toilet Room Accessories **Project No.** ____________

1. Approved submittals, shop drawings, samples, product data, certificates as required. ____________
2. Materials are properly stored on site and protected. ____________
3. All materials and items furnished are approved types and finishes. ____________
4. All handicap accessories are as required and approved. ____________
5. All accessory items, including fastening devices, furnished and are of the approved types. ____________
6. Observe coordination and scheduling of other trades. ____________
7. All templates furnished to other trades, including metal toilet partitions. ____________
8. Toilet room accessories have suitable and secure concealed attachments. ____________
9. Toilet room accessories free from buckle, warping, oil canning, and scratches. ____________
10. All metal-to-metal contacts have hairline joints. ____________
11. All toilet room accessories free from labels, smears, and stains. ____________
12. All toilet accessories firmly and securely attached to surfaces. ____________
13. All toilet accessories cleaned. ____________

General Notes:____

Inspected by:____________________ Date:______________

Figure 1.49 Toilet Room Accessories Checklist.

Section 2

Checklists for Project Management Functions

PROJECT MANAGEMENT AND SITE SUPERVISOR CHECKLISTS

Today's construction projects are complex and the added pressure to meet tight schedules places more demands on project managers and project superintendents. Years ago a million-dollar project was considered meaningful and a ten-million dollar project was considered substantial. Nowadays, projects valued at $20 million and upwards don't raise eyebrows. Along with the increased scope of the construction, more demands are being placed on construction supervisors by added provisions in the construction documents.

Rarely is an AIA (American Institute of Architects) contract executed by contractor and owner by merely filling in the blanks. Teams of lawyers modify the standard provisions in these contracts to the point where 40 or 50 pages of exhibits, exclusions, inclusions, schedules, and modifications place more restrictions on an already restrictive document.

The checklists contained in this section will be especially helpful when mobilizing for new projects. With so many details to attend to such as setting up the field office, reviewing safety requirements, preparing to award contracts for labor, materials, and equipment, these handy "reminders" will prove invaluable.

Jobsite Mobilization Checklist *(Page 1 of 6)*

SITE & SITE SERVICES

1. Site Utilization Plan Prepared
   Approved by: ______________________________
2. Temporary fences, protection (see safety list) ____
3. Guard service ____
4. Temporary electric ____
5. Temporary water ____
6. Dumpster, disposal arrangements ____
   Who pays ______________________________ ____
7. Progress Photograph service
   Who Pays ______________________________ ____
8. Testing Laboratories
   a. Soils ____
   b. Concrete ____
   c. Steel/welding ____
   d. Other __________ ____
   Who Pays ______________________________
9. Weather information phone numbers ____

FIELD OFFICES AND OFFICE EQUIPMENT

1. Facility Type
   a. Trailers ____
   b. Retail Space ____
   c. Space within the construction area ____
   d. Other: __________ ____
2. Number of Offices
   a. Company
      1) Project Manager ____
      2) Superintendent ____
      3) Other: ________ ____
   b. Owner representative(s) ____
   c. Design professionals ____
   d. Subcontractors
      1) __________ ____
      2) __________ ____
      3) __________ ____
      4) __________ ____
      5) __________ ____
      6) __________ ____
3. Temporary Facilities
   a. Heat (Type:_________) ____
   b. Lighting & power ____
   c. Telephones/portable phones
   d. Site communication equipment ____
   e. Lavatories
      1) Use of existing ____
      2) Hookup of Trailers ____
      3) Portable (Quantity:__) ____
   f. Water
      1) Use of existing ____
      2) Hookup to trailers ____

Figure 2.1 Job Mobilization Checklist. *(continued on next page)* *By permission: The McGraw-Hill Company, New York, NY*

Jobsite Mobilization Checklist *(Page 2 of 6)*

4. Office Furniture
   a. Desks, chairs, stools ____
   b. Conference table ____
   c. Folding chairs ____
   d. Plan rack ____
   e. Plan table ____
   f. File cabinets
      1) Regular ____
      2) Locking ____
   g. Book shelves ____
   h. Plan storage cabinets ____
   i. Supply cabinets ____
   j. Plan edge reinforcing machine ____
5. Office Equipment & Supplies
   a. Copier ____
   b. Fax ____
   c. Typewriter
   d. Communication Equipment ____
   e. Computer & Printer ____
   f. Modem ____
   g. Software
      1) Word Processing ____
      2) Scheduling ____
      3) Spreadsheet ____
      4) Database ____
      5) Other: ________ ____
   h. Printing arrangements
      1) Blueprint machine ____
      2) Printing arrangements ____
   i. Refrigerator, coffee machine, supplies ____
   j. Beepers, pagers ____
   k. Copy, computer, fax paper ____
   l. Maintenance items for all equipment ____
6. Office Safety & Security Equipment
   a. Fire & intrusion alarm ____
   b. Fire extinguishers ____
   c. Hard hats
      1) Company personnel ____
      2) Visitors ____
   d. First aid kit & supplies ____
   e. Emergency phone numbers ____
   f. Stretcher ____
   g. Names of any employees with medical training and/or certifications:
      1) ________________________________
      2) ________________________________
      3) ________________________________
      4) ________________________________

Figure 2.1 *(continued from previous page)* Job Mobilization Checklist.

By permission: The McGraw-Hill Company, New York, NY

Jobsite Mobilization Checklist *(Page 3 of 6)*

7. Signs & Notices
   a. Company "Field Office" signs ______
   b. Visitor "Sign-In" Notices ______
   c. Prevailing Wage & other Labor Department notifications ______
   d. "Hard Hat" signs ______
   e. EEO Notices ______
   f. "Keep Out," & "Danger" "Restricted Area" notices ______

ADMINISTRATION

1. Project Manuals & Log Books
   a. Company Operations Manual ______
   b. Project operations Manual ______
   c. Subcontractor Summary & Phone Log ______
   d. Submittal Log ______
   e. Change Order Summary Log ______
   f. RFI Log ______
2. Supply of Job Forms
   a. Daily Field Reports ______
   b. Visitor Sign-In Sheets & Clipboard ______
   c. Change Order forms ______
   d. Quotation & Phone Quote forms ______
   e. Field Payroll forms ______
   f. Administrative Payroll forms ______
   g. Time & Material Tickets ______
   h. Job Meeting forms ______
   i. Record of Meeting/Conversation forms ______
   j. Memos ______
   k. Photograph Record forms ______
   l. Excavation Notification Checklists ______
   m. Equipment Use Release forms ______
   n. Full & Partial Lien Waiver forms ______
   o. Certified Payroll Report forms ______
   p. Transmittal forms ______
   q. Fax Memo/Transmittal forms ______
   r. Field Purchase Order forms ______
   s. Request for Information (RFI) forms ______
   t. Backcharge notices ______
   u. Backcharge forms ______
   v. Schedule status report forms ______
   w. Subtrade Performance Evaluation forms ______
3. Start-Up Project Files
   a. Contract & correspondence files ______
   b. Submittal Files ______
   c. Special Files ______
4. Start-Up Subcontractor Submissions
   a. Subcontracts & Purchase Orders ______
   b. Certificates of Insurance ______
   c. Sub Payment & Performance Bonds ______
   d. Executed Equipment Use Release Forms ______
   e. Approved submittals ______
   f. Other: __________ ______
5. Project Directory ______

Figure 2.1 *(continued from previous page)* Job Mobilization Checklist.

Jobsite Mobilization Checklist *(Page 4 of 6)*

CONSTRUCTION & CONTRACT DOCUMENTS
1. Plans, Specs, Addenda - Construction Set
2. Plans, Specs, Addenda - As-Built Set
3. Project Manual/Working Procedure
4. Building Codes
5. Referenced Standard Specifications
6. Other: ____________

CONTRACT MANAGEMENT DOCUMENTS & GENERAL INFORMATION
1. Owner/Company Agreement
2. Bid Documents
3. Contract Type (GC, CM, CMwGMP, DB)
4. Contract execution date
5. Construction start date
6. Substantial Completion date
7. Final Completion Date
8. Number of Calendar Days
9. Number of Working Days
10. Liquidated Damages
    a. Value/Day ________________
11. Bonus
    a. Value/Day ________________
12. Special Considerations:

CONTRACT EXECUTION
1. Permits — Who Pays — Received
    a. General Building _________
    b. Plumbing _________
    c. HVAC _________
    d. Fire Protection _________
    e. Electrical _________
    f. _______ _________
    g. _______ _________
2. Billing Procedure
    a. Date subcontractor requisitions due
    b. Date general requisition due
    c. Requisition review/approval procedure
3. Change Order Procedure
    a. Change Clause: Section ______
    b. Procedure/forms required
4. EEO Requirements
    a. Mandatory
    b. Good Faith
5. Independent Testing Laboratories
    a. Areas & work types required:
        1) Soils
        2) Concrete
        3) Steel
        4) Other: ___________
    b. Payment responsibility:

Figure 2.1 *(continued from previous page)* Job Mobilization Checklist.

By permission: The McGraw-Hill Company, New York, NY

Jobsite Mobilization Checklist *(Page 5 of 6)*

6. Baselines & Benchmark
   - a. Responsibility to provide: ____
   - b. Responsibility for expense: ____
7. Job Meetings
   - a. Preconstruction meeting date: ____
   - b. Regular meeting schedule: ____
   - c. Who provides official minutes: ____
8. Dispute Resolution
   - a. Dispute resolution clause: Section ____ ____
   - b. Arbitration provision: Section ____ ____
   - c. Notice period:

JOB COST & PRODUCTION CONTROL

1. General project budget ____
2. Resource estimates: material & labor ____
3. Job cost report ____
4. Change Order estimates ____
5. Baseline construction schedule ____
6. Baseline cash-flow projection ____

PROJECT CONTACTS

1. Enforcement Authorities
   - a. General Building Inspector: ____
   - b. Plumbing, HVAC, Fire Protect. Inspector: ____
   - c. Electrical Inspector: ____
   - d. Fire Marshal: ____
   - e. Special Inspectors ____ ____ ____ ____
2. Design Professionals
   - a. Architect: Office Representative: ____
   - b. Architect: Field Representative: ____
   - c. Plumbing, HVAC, FP Engineer(s): ____ ____
   - d. Electrical Engineer: ____
   - c. Structural & Civil Engineer(s): ____ ____
   - d. Landscape Architect: ____
   - e. Special Engineers/Consultants: ____ ____

Figure 2.1 *(continued from previous page)* Job Mobilization Checklist.

By permission: The McGraw-Hill Company, New York, NY

Jobsite Mobilization Checklist *(Page 6 of 6)*

3. Owner Representatives
   a. Field Inspector::
      ____
   b. District Supervisor:
      ____
   c. Other:
      ____
      ____
      ____
4. Security / Life Safety
   a. Police: ______________________________
   b. Fire: ______________________________
   c. Hospital______________________________
   d. Emergency______________________________
   e. Alarm Service______________________________
5. Jobsite Personnel home phone numbers

Figure 2.1 *(continued from previous page)* Job Mobilization Checklist.
By permission: The McGraw-Hill Company, New York, NY

Sample Pre-Start Inspection Checklist

Follow the checklist below to inspect all equipment before your start it - each day that the equipment is used.

1. **Vandalism.** Check to see that:
   a. Smokestacks and exhaust pipes are clear of debris and obstructions. ______
   b. Fuel, water, gas, and oil filters have not been tampered with. ______
   c. Lights and glass are not broken or loosened. ______
   d. Gauges are not damaged. ______
   e. Wires have not been cut. ______
   f. The equipment looks in good overall condition. ______

2. **Tires and Wheels:**
   a. Inspect for loose or missing bolts. ______
   b. Check tire pressure. ______

3. **Operator's Station:**
   a. Be sure it's clean. ______
   b. Check pedals for freedom of movement. ______

4. **Hydraulic System:**
   a. Check Oil Levels ______
   b. Check for leaks, kinked lines, and lines or hoses that rub against each other or other parts. ______

5. **Engine Compartment:**
   a. Check Engine Oil Level. ______
   b. Check Transmission Oil Level. ______
   c. Check Fuel Filter for Sediments. ______
   d. Check Air Cleaner. ______
   e. Check Radiator Coolant Level and clean radiator. ______

6. **Lubrication:**
   a. Check Lubrication Points shown in the respective equipment service manual. ______

7. **Electrical System:**
   a. Check for worn or frayed wires and loose connections. ______

8. **Protective Devices:**
   a. Check guards, canopy, shields, seat belt. ______

9. **Booms, Buckets, Structural Components:**
   a. Check for bent, broken, or missing parts. ______

Figure 2.2 Sample Pre-Start Inspection Checklist. *By permission: The McGraw-Hill Company, New York, NY*

Sample Equipment Service Safety Checklist

1. Put a support under all raised equipment. _____

2. Before beginning service: _____
   a. Review requirements for hard hat, safety shoes, safety glasses or goggles, gloves, reflective vest, ear protectors, and respirator. _____
   b. Be sure that service is approved. _____
   c. Understand all procedures. _____
   d. Stop all equipment. _____
   e. Stop the engine (unless necessary for service). _____

3. Disconnect the battery ground wires before welding or working on the engine or electrical systems. _____

4. Before working on the hydraulic system:
   a. Release all pressure. _____
   b. Loosen fittings slowly. _____

5. Do not smoke:
   a. When you fill the fuel tank. _____
   b. When you work on the fuel system. _____
   c. When you handle fuels or lubricants. _____

6. Do not fill the fuel tank when the engine is running. _____

7. Guard against eye injury when hammering. _____

8. Do not lubricate or work on equipment when it is in motion. _____

9. Avoid high pressure fluids:
   a. Relieve pressure before disconnecting hydraulic or other lines. _____
   b. Tighten all connections before applying pressure. _____
   c. Keep hands and body away from pinholes and nozzles that eject fluids under high pressure. _____
   d. Use a piece of cardboard to search for leaks - do not use your hand. _____

10. Engine Coolant:
    a. Only add coolant to the radiator when the engine is stopped or running at slow idle. _____
    b. Do not remove cap unless engine is cool. _____
    c. Release all pressure before removing cap, and loosen slowly. _____

11. Tires:
    a. Do not attempt to mount tires without the proper equipment and experience to perform the job safely. Failure to follow proper procedure can produce an explosion which may result in serious bodily injury or death. _____
    b. Be sure all tire rims are correctly assembled and interlocking before inflating tires.
    c. Use an inflation cage, safety cables, or other protective device during inflation. _____

Figure 2.3 Sample Equipment Service Safety Checklist. *By permission: The McGraw-Hill Company, New York, NY*

Sample General Equipment Safety Checklist

1. No persons other than company personnel are allowed to operate any company equipment under any circumstances. ____
2. Only qualified operators are allowed to operate the equipment. ____
3. Learn the location and purpose of all controls, instruments, indicator light, and labels. ____
4. Be sure that a first aid kit is fastened to all major equipment. ____
5. Keep a fully charged fire extinguisher mounted conveniently. Learn to use it correctly. ____
6. Wear fairly tight, properly fitting clothing and safety equipment. ____
7. Avoid high-pressure fluids that can penetrate skin and cause injury. ____
8. Relieve pressure before disconnecting hydraulic or other lines. ____
9. Tighten all connections before applying pressure. ____
10. Keep hands and body away from pinholes and nozzles that eject fluids under high pressure. ____
11. Use a piece of cardboard to search for leaks. Do not use your hand. ____
12. ***If ANY fluid is ejected into the skin, it must be surgically removed within a few hours by a doctor or gangrene may result.*** ____
13. Wear a suitable hearing protective device such as earmuffs or earplugs, in order to protect against loud noise. ____

Figure 2.4 Sample General Equipment Safety Checklist. *By permission: The McGraw-Hill Company, New York, NY*

Sample Equipment Fire Prevention Maintenance Checklist

1. Daily Pre-Start Maintenance:
   a. Check fire extinguisher for correct charge. ____
   b. Open all access hoods and shields.
      Remove all trash from all areas inside these compartments from:
      i. Exhaust manifold, turbocharger, and muffler. ____
      ii. Bottom guards and under engine. ____
      iii. Sides of engine. ____
      iv. Radiator and oil cooler. ____
      v. Batteries. ____
      vi. Hydraulic lines. ____
      vii. Fuel tank. ____
2. Check for leaking fuel lines, hydraulic lines, or fittings.
   a. Tighten loose fittings. ____
   b. Replace bent or kinked lines. ____
3. Clean trash from grilles. ____
4. Clean trash from cab areas. ____
5. Be sure all doors and grilles are in place. ____
6. Shut-Down:
   a. Be on guard for fires; especially when refueling.
      Temperature in engine compartment may go up immediately after stopping engine. ____
   b. Wait until the engine has cooled before filling the fuel tank. ____
   c. Do not smoke when refueling. ____

Figure 2.5 Sample Equipment Fire Prevention Maintenance Checklist.
By permission: The McGraw-Hill Company, New York, NY

Sample Equipment Safe Operation Checklist

1. Use handrails and steps to enter and leave the operator's station. Do not use the steering wheel. _____
2. Keep handrails, steps, floor, and controls free of water, grease, and dirt. _____
3. Do not operate any equipment in an unsafe condition. Put a tag on the steering wheel or other appropriate high-visibilty location. _____
4. Before starting or operating any equipment:
   a. Check the condition of the equipment (See Pre-Start Inspection Checklist below). _____
   b. Be sure there is enough ventilation. _____
   c. Know the correct starting and stopping procedure. _____
   d. Sit in the operator's seat. _____
   e. Clear the work area of people and obstacles. _____
   f. Check the service brakes and parking brake _____

7. Be sure engine is running and foot brakes are operating before releasing the parking brake. _____
8. Do not allow riders on the equipment. _____
9. Drive slowly in congested areas, over rough ground, and on slopes and curves. _____
10. Do not drive near the edges of ditches and excavations. _____
11. Keep loading areas smooth. _____
12. Check locations of utilities, cables, gas lines, water mains, etc. before digging. _____
13. Keep away from power lines at all costs. Do not touch power lines with any part of the equipment. _____
14. Carry buckets and loads as low as possible for better stability and visibility. _____
15. Keep equipment in gear when going down steep grades. _____
16. Use accessory lights and devices to warn operators of other vehicles. _____
17. Position backhoe booms on uphill side when driving across hillsides. _____
18. Set stabilizers before operating any backhoe equipment. _____
19. Use care when raising stabilizers; they may be the only restraint preventing stability. _____
20. Do not dig under stabilizers. _____
21. Avoid swinging any backhoe bucket in the downhill direction. _____
22. Before dismounting the equipment:
   a. Engage parking brake. _____
   b. Lower all equipment to the ground. _____
   c. Stop engine. _____
   d. Release hydraulic pressure; turn steering wheel back and forth, move hydraulic control levers until the equipment does not move. _____

Figure 2.6 Sample Equipment Safe Operation Checklist. *By permission: The McGraw-Hill Company, New York, NY*

Sample Winter Precautions Checklist *(Page 1 of 4)*

		YES	NO
A.	**GENERAL PROJECT STATUS**		
	As of ______________________		
1.	Building portions satisfactorily closed to weather:		
	a. Roofs & Flashings	___	___
	b. Doors & Windows	___	___
	c. Building Skin	___	___
	d. ______________________	___	___
	e. ______________________	___	___
2.	Permanent heating system usable for temporary heat:		
	a. Electrical	___	___
	b. HVAC	___	___
3.	Interior pipes/systems subject to freezing:		
	a. Remarks: ______________________		
4.	Permanent Source of Power Available:	___	___
5.	Temporary power Necessary:	___	___
	a. Remarks: ______________________		
6.	Permanent Source of Fuel Available:	___	___
7.	Temporary Fuel Necessary:	___	___
B.	**CONTRACT ASSESSMENT**		
1.	Temporary heat required between (dates) ____________ and ____________		
2.	Responsibility to provide temporary heat:		
	a. Owner:	___	___
	b. Prime Contractor or Construction Manager:	___	___
	c. Sub or Trade Contractor(s):		
	(1) ______________________	___	___
	(2) ______________________	___	___
	(3) ______________________	___	___
3.	Responsibility to provide temporary protection:		
	a. Owner:	___	___
	b. Prime Contractor or Construction Manager:	___	___
	c. Sub or Trade Contractor(s):		
	(1) ______________________	___	___
	(2) ______________________	___	___
	(3) ______________________	___	___

Figure 2.7 Sample Winter Precautions Checklist. *(continued on next page)*

Sample Winter Precautions Checklist *(Page 2 of 4)*

		YES	NO
4.	Temporary heat now required because of delay:	___	___
5.	If (4) yes, who is responsible:		
	a. Owner:	___	___
	b. Prime Contractor or Construction Manager:	___	___
	c. Sub or Trade Contractor(s):		
	(1) ____________________	___	___
	(2) ____________________	___	___
	(3) ____________________	___	___
	d. Reasons/Remarks: ____________________		
6.	If (5) is Owner:		
	a. Change Order File established	___	___
	b. Change Order acknowledged by the Owner	___	___
	c. Is a claim necessary (denied Change Order)	___	___
	d. If (6.c.) yes:		
	(1) Written notification made Date: ____________________ To: ____________________	___	___
	(2) Documentation provided ____________________	___	___
7.	If (5) is Subcontractor or Trade Contractor:		
	a. Has backcharge procedure begun	___	___
	b. Written Backcharge Notice sent	___	___
	c. Responsibility accepted	___	___
8.	Estimated cost of temporary services (attached detailed estimate forms)		
	a. Protection $ ____________________		
	b. Heating Equipment $ ____________________		
	c. Heating Fuel $ ____________________		
	d. Light & Power $ ____________________		
	e. Total $ ____________________		

Figure 2.7 *(continued from previous page)* Sample Winter Precautions Checklist.

By permission: The McGraw-Hill Company, New York, NY

Sample Winter Precautions Checklist *(Page 3 of 4)*

		YES	NO
C.	**OVERALL JOB PRECAUTIONS**		
1.	Arrangements made to secure:		
	a. Temporary protection materials	___	___
	b. Temporary enclosure materials	___	___
	c. Continuous fuel supply	___	___
2.	Temporary heating equipment is:		
	a. Of adequate size & type	___	___
	b. Is maintained and fully operational	___	___
	c. Of type(s) allowed by codes	___	___
	d. Situated in safe manner relative to pedestrians, traffic, building materials, and ventilation	___	___
	e. On the service/maintenance schedule	___	___
3.	Temporary fuel is:		
	a. On hand and in adequate supply	___	___
	b. Properly and safely stored	___	___
	c. On a set refueling schedule	___	___
4.	All water pockets have been eliminated:		
	a. Roof areas	___	___
	b. Pavement and graded areas	___	___
	c. Sleeves, inserts, chases & openings	___	___
	d. Other: ______________________________	___	___
5.	Arrangements have been made for:		
	a. Snow plowing/removal	___	___
	b. Equipment cold weather protection	___	___
	c. Vehicle Maintenance	___	___
6.	Precautions taken to protect exposed work:		
	a. Exposed piping protected, drained, or heat traced	___	___
	b. Recently placed work (concrete, form work, reinf. steel, masonry, etc.)	___	___
7.	All project areas have been adequately marked to avoid damage during snow removal:		
	a. Parking areas	___	___
	b. Entrances, exits, gates, passageways	___	___
	c. Pedestrian traffic areas	___	___
	d. Materially and fuel storage areas	___	___
8.	Any necessary photographs of all pre-winter jobsite conditions taken for record	___	___

Figure 2.7 *(continued from previous page)* Sample Winter Precautions Checklist.

Sample Winter Precautions Checklist *(Page 4 of 4)*

D. SPECIFIC WINTER PRECAUTIONS

1. **Item of work**: ______________________
   Location: ______________________
   ____
   Party Responsible: ______________________
   Specific Precautions Taken: ______________________
   ____
   ____
   ____
   Precaution Start Date: ____________
   Anticipated End Date: ____________
   Remarks: ______________________
   ____

2. **Item of work:** ______________________
   Location: ______________________
   ____
   Party Responsible: ______________________
   Specific Precautions Taken: ______________________
   ____
   ____
   ____
   Precaution Start Date: ____________
   Anticipated End Date: ____________
   Remarks: ______________________
   ____

3. **Item of work:** ______________________
   Location: ______________________
   ____
   Party Responsible: ______________________
   Specific Precautions Taken: ______________________
   ____
   ____
   ____
   Precaution Start Date: ____________
   Anticipated End Date: ____________
   Remarks: ______________________
   ____

Figure 2.7 *(continued from previous page)* Sample Winter Precautions Checklist.

Sample Jobsite Safety Review Checklist (Page 1 of 5)

1. Signs, Notices, & Notifications
   a. Safety signs in place ____
   b. Emergency phone numbers posted ____
   c. Evacuation plan approved/posted ____
   d. Warning & instructions to public posted ____
   e. Respected access areas ____
   f. Exits ____
   g. No Smoking ____
   h. Electrical dangers ____
   i. Personal protective equipment needed ____
   j. Operating instructions ____
   k. Flammable materials ____
   l. Hazardous materials ____
   m. Danger areas ____
   n. Trenches ____
   o. All personnel & occupants notified to expect loud noises ____
   p. ________________ ____
   q. ________________ ____

2. Overhead Protection
   a. Entrances ____
   b. Warnings ____
   c. Construction ____
   d. ________________ ____
   e. ________________ ____

3. Hoisting Equipment
   a. Guys ____
   b. Cables & Sheaves ____
   c. Turnbuckles ____
   d. Signals ____
   e. Car cover & enclosure ____
   f. Ladder ____
   g. Car arresting device ____
   h. Base barricade ____
   i. Platforms ____
   j. Clear staging areas ____
   k. ________________ ____
   l. ________________ ____

4. Walkways & Ramps
   a. Adequate construction ____
   b. Width ____
   c. Railings ____
   d. Curbs ____
   e. Slope & rise limit ____
   f. Non-slip treads & tactile areas ____
   g. ________________ ____
   h. ________________ ____

Figure 2.8 Sample Jobsite Safety Review Checklist. *(continued on next page)*

Sample Jobsite Safety Review Checklist (Page 2 of 5)

5. Ladders
   a. Construction ____
   b. Secure placement ____
   c. Cleats ____
   d. Landings ____
   e. Hand-holds ____
   f. Cages ____
   g. ____________________ ____
   h. ____________________ ____

6. Excavations & Trenches
   a. Shoring ____
   b. Slope repose ____
   c. Ladders ____
   d. Stockpile of excavated material ____
   e. Removal of excavated material ____
   f. Barricades & railings ____
   g. Tunnels ____
   h. Blasting arrangements ____
   i. Approved shoring designs ____
   j. Excavations properly dewatered ____
   k. Proper ventilation; free of toxic fumes ____
   l. ____________________ ____
   m. ____________________ ____

7. Fire Protection
   a. Storage of flammable materials ____
   b. Container markings ____
   c. Temporary heaters ____
   d. Compressed gas cylinders ____
   e. Tar kettles ____
   f. Welding equipment ____
   g. Welding operations ____
   h. Fire extinguishers (correct quantities/type) ____
   i. Fire safety equipment ____
   j. ____________________ ____
   k. ____________________ ____

8. Openings - Walks, Floors, Roofs
   a. Perimeter railings ____
   b. Tight covers ____
   c. Flaggings ____
   d. ____________________ ____
   e. ____________________ ____

Figure 2.8 *(continued from previous page)* Sample Jobsite Safety Review Checklist.

Sample Jobsite Safety Review Checklist (Page 3 of 5)

9. Scaffolds
   a. Construction _____
   b. Secure placement _____
   c. Railings _____
   d. Toe boards _____
   e. Rigging _____
   f. Safety lines, belts, rope guards _____
   g. ____________________ _____
   h. ____________________ _____

10. Stairs & Landings
   a. Adequate construction _____
   b. Temporary treads _____
   c. Clear of debris _____
   d. Proper rise/run _____
   e. Railings _____
   f. ____________________ _____
   g. ____________________ _____

11. Material Handling
   a. Size/Bulk _____
   b. No sharp edges _____
   c. Weight limits _____
   d. Team lifting _____
   e. ____________________ _____
   f. ____________________ _____

12. Housekeeping
   a. Nails, debris _____
   b. Tool storage and staging _____
   c. Containers _____
   d. Clear aisles & walkways _____
   e. Clean site _____
   f. Dumpster(s) location/condition _____
   g. Proximity of waste storage to hazardous conditions _____
   h. ____________________ _____
   i. ____________________ _____

13. Lighting & Temporary Wiring
   a. Lighting _____
   b. Wire height _____
   c. Proper grounding _____
   d. Wire connection _____
   e. Overcurrent protection _____
   f. Extension chords in good repair _____
   g. All extension chords & temporary power receptacles using GFIs _____
   h. Temporary power closed to weather _____
   i. ____________________ _____
   j. ____________________ _____

Figure 2.8 *(continued from previous page)* Sample Jobsite Safety Review Checklist.

Sample Jobsite Safety Review Checklist (Page 4 of 5)

14. Grounding & Electrical Equipment
    a. Correct Grounding ____
    b. Ground Fault Interrupters ____
    c. ________________________ ____
    d. ________________________ ____

15. Portable & Power Saws
    a. In good condition ____
    b. Guards ____
    c. Kickback protection ____
    d. Ventilation ____
    e. Safe fuel procedures ____
    f. ________________________ ____
    g. ________________________ ____

16. Hand Tools
    a. In good condition ____
    b. Insulated and/or grounded ____
    c. Projectile tools ____
    d. Power actuated tools ____
    e. Operators trained in proper use ____
    f. ________________________ ____
    g. ________________________ ____

17. First Aid
    a. Proper kit size & contents ____
    b. Kit supply maintained ____
    c. Trained employees ____
    d. Emergency numbers posted ____
    e. Hospital & emergency routes known ____
    f. ________________________ ____
    g. ________________________ ____

18. Traffic Control
    a. Parking ____
    b. Speed control ____
    c. Barricades ____
    d. Separation of haul roads ____
    e. ________________________ ____
    f. ________________________ ____

19. Personal protective equipment
    a. Hard hats ____
    b. Goggles / safety glasses ____
    c. Gloves ____
    d. Respirators ____
    e. Hearing protection ____
    f. Safety shoes ____
    g. No loose clothing ____
    h. All work areas sanitary ____
    i. ________________________ ____
    j. ________________________ ____

Figure 2.8 *(continued from previous page)* Sample Jobsite Safety Review Checklist.

By permission: The McGraw-Hill Company, New York, NY

Sample Jobsite Safety Review Checklist (Page 5 of 5)

20. Heavy Equipment
    a. Guards ____
    b. Warning bells ____
    c. Fueling ____
    d. Ground slope ____
    e. Rough terrain ____
    f. Cab Protection ____
    g. Operator qualifications ____
    h. ____________________ ____
    i. ____________________ ____

21. Security
    a. Fencing ____
    b. Lighting ____
    c. Alarm systems ____
    d. Monitoring arrangements ____
    e. Guard service ____
    f. Target equipment ____
    g. Secure equipment practices ____
    h. Police notification procedure ____
    i. ____________________ ____
    j. ____________________ ____

22. Liability
    a. Release forms executed / delivered for all trades using:
        (1) Hoists ____
        (2) Elevators ____
        (3) Scaffolding ____
        (4) Equipment ____
    b. Arrange for jobsite inspection by insurance carrier ____
    c. ____________________ ____
    d. ____________________ ____

23. Other:
    a. ____________________ ____
    b. ____________________ ____
    c. ____________________ ____
    d. ____________________ ____
    e. ____________________ ____
    f. ____________________ ____
    g. ____________________ ____
    h. ____________________ ____
    i. ____________________ ____
    j. ____________________ ____
    k. ____________________ ____
    l. ____________________ ____
    m. ____________________ ____
    n. ____________________ ____
    o. ____________________ ____

Figure 2.8 *(continued from previous page)* Sample Jobsite Safety Review Checklist.
By permission: The McGraw-Hill Company, New York, NY

Sample Jobsite Safety Planning Checklist

(Page 1 of 3)

1. Administration: Establish and maintain on-file adequate supplies of:
   a. Site Safety Program ____
   b. Safety Review Checklists ____
   c. Posted listings of emergency services ____
   d. Accident Investigation Report Forms ____
   e. Accident Eyewitness Statement Outline ____
   f. Tailgate Safety Meeting Agendas ____
   g. OSHA forms 101 & 200 ____
2. Forward "Job Startup Notice" to company Insurance Carrier ____
3. Identify all formal and available informal Safety Personnel:
   a. Company Safety Officer ____
   b. Company On-Site Safety Representative ____
   c. Owner Safety Representative ____
   d. All company employees with first aid, CPR or other safety training ____
   e. All Owner employees with first aid, CPR or other safety training ____
   f. All Sub-Vendor employees with first aid, CPR or other safety training ____
4. Identify Safety & Emergency Services:
   a. Notify of the presence of the jobsite, anticipated work force, and duration of the project:
      i. Fire Department ____
      ii. Police Department ____
      iii. Medical Facilities ____
      iv. Identify locations and emergency travel routes to:
         (a) Emergency rooms ____
         (b) Non-emergency medical facilities ____
         (c) Fire Department ____
         (d) Police Department ____
5. Determine Listings of Local Services:
   a. Physicians ____
   b. Eye Experts ____
   c. Orthopedics ____
   d. Paramedics ____
6. Unique Owner Requirements (In addition to "usual" considerations)
   a. Construction Parking ____
   b. Occupied Area Parking ____
   c. Strict Security ____
   d. Hot Work or other Special Work Permits ____
   e. Full or Part-Time Safety Person ____
   f. Insurance wrap-up policy ____

Figure 2.9 Sample Jobsite Safety Planning Checklist. *(continued on next page)*

Sample Jobsite Safety Planning Checklist

(Page 2 of 3)

7. Traffic Control - a plan to address:
   a. Entering/leaving of the construction work force ____
   b. Material/Equipment Deliveries ____
   c. Equipment access & storage ____
   d. Street & site cleanup & maintenance ____
   e. Adequate warning signs & other postings ____

8. Utility Location Protection - Locate & Protect/Relocate:
   a. Water ____
   b. Gas ____
   c. Electric ____
   d. Telephone ____
   e. Telex/Cable ____
   f. Communication ____
   g. Storm/Sanitary sewer ____

9. Preexisting Condition Survey

   Prior to the start of any construction, record all preexisting damage by whatever means appropriate, including engineered surveys, photographs, or videotape recordings.

   a. Initial survey should be made by the Site Superintendent ____
   b. Determine the need for additional detailed inspection(s). ____
   c. If sufficient evidence of damage is present, consider retaining a professional photographer to record it. ____
   d. If extreme or unique circumstances, determine if a professional engineer is or may be needed to produce surveys or other documentation. ____
   e. Prior to proceeding with (c) or (d), secure approval from the company Senior Project Manager. ____
   f. In every case, photograph and videotape the entire site, as well as all approach roads and routes prior to site mobilization and start of any construction activity, Include:
      i. Surrounding Buildings ____
      ii. Roads ____
      iii. Utilities ____

10. Plan the locations and configurations of site services and their relative proximities:
    a. Field Offices ____
    b. Material Staging and storage areas ____
    c. Fuel Storage and fuel distribution arrangements ____
    d. Traffic Control ____
    e. Administrative and worker parking ____
    f. Pedestrian access ____
    g. Location & configuration of temporary utilities ____
    h. Ongoing temporary power arrangement ____
    i. Temporary lighting - Site ____
    j. Temporary lighting - Other Construction Areas ____
    k. Temporary heat ____
    l. Temporary Power ____
    m. Welding & cutting torches (friendly fire) ____

Figure 2.9 *(continued from previous page)* Sample Jobsite Safety Planning Checklist.

By permission: The McGraw-Hill Company, New York, NY

Sample Jobsite Safety Planning Checklist

(Page 3 of 3)

11. Identify specific Owner requirements that may exceed customary considerations, such as:
    a. Special access requirements or needs ____
    b. Unique security requirements ____
    c. Involvement of Owner or other designated safety or security personnel ____
    d. Special forms of insurance or legal considerations ____
    e. Special Notices or other required communications ____

12. Determine necessary provisions for protection of the public:
    a. Site fencing ____
    b. Lighting ____
    c. Signs & Notices ____
    d. Traffic Control ____
    e. Guardrails ____
    f. Walkways (Covered/Uncovered) ____

13. Determine all Fire Protection needs - Project-specific planning should include:
    a. Appropriate ABC-type fire extinguishers:
       i. Sizes ____
       ii. Quantities ____
       iii. Locations ____
    b. Fire-protected storage cabinets and/or areas for vital, sensitive, or special files or materials ____
    c. Establishment and maintenance of all appropriate fire protection measures in accordance with OSHA requirements for specific work operations ____

14. Consider necessary protection of the Site and Building (if there is one)
    a. Signs and notices posted ____
    b. Special walkways and traffic provisions ____
    c. Barricades and other safety barriers ____
    d. Fences, guardrails, and canopies ____
    e. Parking, traffic & walkway lighting ____
    f. All provisions for handicap access ____
    g. Security & safety personnel ____

Figure 2.9 *(continued from previous page)* Sample Jobsite Safety Planning Checklist.

Submittal Review Checklist *(Page 1 of 2)*

SUBMISSION REQUIREMENTS

All approvals/submissions contain:

1. Project Title and Job Number ____
2. Contract Identification ____
3. Date of Submission (or Revision) ____
4. Dates of Previous Submissions ____
5. Names of contractor, supplier and/or manufacturer ____
6. Identification of all products with specification section numbers ____
7. Field dimensions - clearly identified as such ____
8. Relation to adjacent and/or critical features of the work ____
9. References to applicable standard specifications ____
10. *Clear* identification of deviations from the Contract Documents ____
11. All other pertinent information as may be required by the specifications with the Company, such as:
    a. Model Numbers ____
    b. Performance Characteristics ____
    c. Dimensions & Clearances ____
    d. Wiring or Piping Diagrams ____
12. Manufacturer's standard drawings include:
    a. Modifications to delete information not applicable to this project ____
    b. Supplemental information specifically applicable to this project ____
13. Check the specifications for additional requirements ____

SUBMITTAL REVIEW PROCEDURE

1. Ensure that subcontractors and suppliers submit materials promptly ____
2. Determining and verify:
    a. That the sub has incorporated and will guarantee all field dimensions ____
    b. All field conditions and construction criteria have been accommodated ____
    c. That the product either complies with the specification requirements in every respect, or that every deviation has been properly identified, and includes its respective complete explanation/justification ____
3. Coordinate each submittal with all field and contract document requirements ____
4. Research and confirm all "justifications" for any deviation from the Contract Documents. Do this *before* submitting the documents for approval. ____
5. Determine if a credit or addition to the contract is in order, based upon any changes in the submission ____
6. Determine if any backcharges to any other subcontractor or supplier are in order as a result of changes required by this item ____
7. Determine that the submission is timely, and that the material conforms to all delivery requirements ____
8. Positively identify by responsibility all "Not by Subcontractor" or "By Others" kinds of remarks before submission to the architect for approval. ____
9. Compare all re-submissions with the file copy of the previous submission. Confirm that all required corrections have been made and no new issues have been introduced ____

Figure 2.10 Submittal Review Checklist. *(continued on next page)* *By permission: The McGraw-Hill Company, New York, NY*

Submittal Review Checklist *(Page 2 of 2)*

DISTRIBUTION

1. Upon receipt of submittals bearing the stamp indicating architect action, distribute copies to:
   a. Jobsite File ("For Construction" documents only) ____
   b. Record Documents File ____
   c. Other affected subcontractors and suppliers ____
   d. The supplier or fabricator ____
   e. Anyone else who may need information in order to coordinate the work ____

FOLLOW-UP

1. Monitor the time that it takes for the approval process:
   a. Be sure that the architect is giving proper, timely attention ____
   b. Be sure that all delays and other inappropriate action are duly noted in the correspondence ____
2. Be certain that the design professionals:
   a. Include all information required of them by way of questions in the submittals ____
   b. Do not overstep their authority ____
   c. Do not overstep their professional capacities ____
   d. Do not add work without regard for the established Change Order procedure ____
   e. Include only meaningful action that will allow proper completion of the submittal ____
   f. Affix the *accept* stamped and initial/sign it ____
   g. Clearly indicate any requirements for resubmittal, or approval of the submittal ____
3. Upon distribution of the submittal back to its originator:
   a. Reconfirm the delivery schedule(s) ____
   b. Confirm that the submission is being returned in good time for the subcontractor or supplier to meet its own requirements ____
   c. Note any significant information for the next construction schedule update ____
   d. Begin any actions that may be necessary to resolve problems that may have been exposed by the review process ____
   e. Begin any necessary Change Order procedure ____
4. Be sure that the Submittal Log form is used in maintaining *as each part of the process is completed*. Complete the respective Log entry ____

Figure 2.10 *(continued from previous page)* Submittal Review Checklist.

By permission: The McGraw-Hill Company, New York, NY

Sample Project Closeout Checklist *(Page 1 of 3)*

1. All systems in operation:
   a. Plumbing ____
   b. HVAC ____
   c. Electrical ____
   d. Fire Protection ____
   e. Conveying Systems ____
   f. Communications ____
   g. Building Management Systems ____
   h. ________________________ ____
   i. ________________________ ____

2. Performance tests conducted:
   a. Plumbing ____
   b. HVAC ____
   c. Electrical ____
   d. Fire Protection ____
   e. Conveying Systems ____
   f. Communications ____
   g. Building Management Systems ____
   h. ________________________ ____
   i. ________________________ ____

3. O & M Manuals delivered:
   a. Plumbing ____
   b. HVAC ____
   c. Electrical ____
   d. Fire Protection ____
   e. Conveying Systems ____
   f. Communications ____
   g. Building Management Systems ____
   h. ________________________ ____
   i. ________________________ ____

4. Operating Instructions to Owner performed::
   a. Plumbing ____
   b. HVAC ____
   c. Electrical ____
   d. Fire Protection ____
   e. Conveying Systems ____
   f. Communications ____
   g. Building Management Systems ____
   h. ________________________ ____
   i. ________________________ ____

5. Final completion of physical work
   a. Company Punchlist complete ____
   b. Architect/Owner Punchlist complete ____
   c. All sign-off forms completed ____
   d. Completion Certificate(s) received ____

Figure 2.11 Sample Project Closeout Checklist. *(continued on next page)*

By permission: The McGraw-Hill Company, New York, NY

Sample Project Closeout Checklist *(Page 2 of 3)*

6. Demobilization complete:
   a. Field Offices ____
   b. Equipment & furnishings ____
   c. ________________________ ____
   d. ________________________ ____

7. Termination of Temporary Services
   a. Heat, light, power, phone ____
   b. Fire, police, guard service ____
   c. Insurance transfers ____
   d. ________________________ ____
   e. ________________________ ____

8. Final cleaning completed
   a. ________________________ ____
   b. ________________________ ____
   c. ________________________ ____
   d. ________________________ ____

9. As-Built Drawings
   a. General ____
   b. Plumbing ____
   c. HVAC ____
   d. Electrical ____
   e. Fire Protection ____
   f. Conveying Systems ____
   g. Communications ____
   h. Building Management Systems ____
   i. ________________________ ____
   j. ________________________ ____

10. Guarantees & Warranties
   a. General ____
   b. All subcontractor documents ____
   c. Special bonds and other third-party documents ____
   d. ________________________ ____
   e. ________________________ ____

11. Inspection Certificates
   a. ________________________ ____
   b. ________________________ ____
   c. ________________________ ____
   d. ________________________ ____

Figure 2.11 *(continued from previous page)* Sample Project Closeout Checklist.

By permission: The McGraw-Hill Company, New York, NY

Sample Project Closeout Checklist *(Page 3 of 3)*

12. Material/Installation Certificates
    a. ______________________________ ____
    b. ______________________________ ____
    c. ______________________________ ____
    d. ______________________________ ____

13. Lien Waivers & General Release Forms
    a. General ____
    b. All Subcontractor Documents ____
    c. ______________________________ ____
    d. ______________________________ ____

14. Billing & Charges processed
    a. All acknowledged Change Orders ____
    b. All subcontractor changes ____
    c. All subcontractor backcharges ____
    d. Final billings from Sub-Vendors ____
    e. Final billing submitted to the Owner ____

15. Steps taken to finalize outstanding claims
    a. To Owner ____
    b. To Subcontractors & Suppliers ____
    c. By Subcontractors ____
    d. ______________________________ ____
    e. ______________________________ ____

16. Project Completion Report submitted ____

17. Project Records transferred to home office ____

18. Forwarding address confirmed
    a. Post Office notified ____
    b. All subcontractors and suppliers notified ____
    c. ______________________________ ____

19. All other in general contractor, subcontractor, purchase order, specification, and Company Procedure items necessary to close-out the project:
    a. Attach list ____
    b. ______________________________ ____
    c. ______________________________ ____
    d. ______________________________ ____

Figure 2.11 *(continued from previous page)* Sample Project Closeout Checklist.
By permission: The McGraw-Hill Company, New York, NY

Sample Checklist: Factors Affecting Labor Productivity

(Page 1 of 2)

ITEM		Relative % of Loss: Minor	Average	Severe
1.	**STACKING OF TRADES:** Operations can take place within physically limited space that a shared with other contractors. This can result in congestion of personnel, inability to locate tools conveniently, increased tool loss-rate, increase of additional safety hazards, and added increased visitors. Optimum crew sizes in many cases cannot be utilized. Scheduling can become haphazard and inefficient. Idle time and down-time increases.	5%	15%	25%
2.	**LABOR REASSIGNMENT:** Unnecessarily high levels of labor movement on and off projects due to scheduling, unexpected changes, excessive changes, demands made to expedite production to make up for performance deficiencies of others, and unnecessary rework due to over-inspection or unfounded criticisms of quality. Orderly changes interfered with or otherwise not possible.	5%	10%	15%
3.	**MORALE AND ATTITUDE:** Cooperation among contractors, poor supervision from the prime contractor, over-inspection, poor coordination, multiple changes, unfounded criticisms of quality, disruption of pace, poor housekeeping, and ineffective material management can all create or increase hazards. Competition for overtime and other personnel matters can decrease cooperation among employees.	5%	10%	20%
4.	**DILUTION OF SUPERVISION:** Increased number of work activities caused by improper planning and increased number of unexpected change activities, unplanned starting and stopping of activities, the need to change material scope and quantities, receiving incomplete or intermittent punchlists, inability to properly coordinate testing or start-up activities, and other items that cause supervision to be spread thin.	5%	15%	20%
5.	**LEARNING CURVE:** Some learning curve may be expected at the beginning of a project or task, but may become extended due to unplanned start-stop or other interruptions caused by factors beyond our control. This factor is dramatically impacted by labor reassignment and high rates of crew turnover.	5%	15%	25%
6.	**CREW SIZE INEFFICIENCY:** Adding workers to optimally-sized crews can interfere with orderly work and affect the rhythm of the labor output. After a point, added labor can become very inefficient; dramatically and adversely affecting any planned levels of productivity.	3%	8%	15%
7.	**FATIGUE:** Unusual physical exertion, including any manner of excessive overtime hours. Ongoing 6 and 7-day work weeks. Jumping between contract and change order work, and continually working in a poorly coordinated environment increases physical fatigue factors.	5%	8%	15%
8.	**CONCURRENT OPERATIONS:** Unreasonable stacking of the workforce, including the effects of adding additional operations to those already planned in orderly sequence. Sudden and uncontrolled implementation of additional operations, instead of orderly and gradual increases.	5%	15%	20%
9.	**SITE ACCESS:** Prevention of access to key areas. Interference with convenient regular access to all needed areas not just for work areas, but for personnel-lift and equipment logistics arrangements as well. Congested site. Poor housekeeping and poor material waste management. Inability to use planned equipment.	5%	15%	25%
10.	**LOGISTICS:** Problems arising with material storage and staging. Uncoordinated and intermittent delivery of materials and equipment furnished by others. Little or no control over material flow to work areas. Problems caused by changes that affect procurement and delivery of materials. Re-handling of substituted materials.	5%	10%	20%
11.	**RIPPLE EFFECTS:** Impact on our work caused by delays or other problems to the work of other trades. Changes that affect other items of work or work areas that prevent the orderly progression of our own work.	5%	10%	15%

Figure 2.12 Sample Checklist: Factors Affecting Labor Productivity. *(continued on next page)*

Sample Checklist: Factors Affecting Labor Productivity

(Page 2 of 2)

ITEM		Relative % of Loss: Minor	Average	Severe
12.	**EXCESSIVE OVERTIME:** Labor output and work efficiency is lowered due to physical fatigue, dilution of supervision, concurrent operations, crew size inefficiency, ripple effects and worsening attitudes	10%	18%	50%
13.	**SEASONAL AND WEATHER CHANGES:** Unexpected occurrence of weather extremes, including heat, cold, rain, snow, and wind. Effects are compounded if operations are forced into adverse weather due to delays or other interferences that are beyond our control.	5%	10%	15%
14.	**DESIGN OR OTHER INSTRUCTIONAL ERRORS AND OMISSIONS:** Work interruptions caused by late discovery of design errors or mistakes in instruction. Changes performed on an expedited, out-of-sequence basis, with problems contributed by dilution of supervision, excessive overtime, learning curve, and crew size inefficiency.	5%	8%	10%
15.	**BENEFICIAL OCCUPANCY:** The need to work around, over, and/or in close proximity to owner's personnel, furnishings, and equipment. Increased access issues such as parking, dust protection, noise limitations, increased safety activities, and limited hours of work. Continued use of work areas conditioned on some system of notification and prior arrangement.	15%	25%	35%
16.	**JOINT OCCUPANCY:** Continuing work beyond our originally-anticipated time period while the work area is occupied by other trades that were not anticipated under the original bid conditions.	5%	10%	15%

Figure 2.12 *(continued from previous page)* Sample Checklist: Factors Affecting Labor Productivity.
By permission: The McGraw-Hill Company, New York, NY

Telephone Quote Checklist

(Page 1 of 2)

1. Verify all specific information

   a. Specific material:
      i. "As specified" ?
      ii. "As equal?"
      iii. "As a substitution?"

   b. Price:
      i. Lump sum?
      ii. Unit price?
      iii. Applicable taxes?
      iv. Other?

   c. Price Incentives:
      i. Discounts for combined or consolidated orders?
      ii. Other conditions for reduced costs?

   d. Submittals:
      i. Per plans and specifications (form and content)?
      ii. Number of copies?
      iii. Date of submittal delivery (In total? In parts?)

   e. Material Delivery(ies):
      i. Time after receipt of approved submittals?
      ii. Single shipment or broken up?
      iii. Shop inspection availability (Required? Advisable?)

   f. Acceptance of General Contract Terms and Conditions:
      i. No Exceptions?
      ii. Acceptable Exceptions?
      iii. Unacceptable Exceptions?

   g. Payment terms:
      i. Pay-when-paid?
      ii. Pay-if-Paid?
      iii. Retainage to be withheld? If so, what percentage?
      iv. "Net" payment terms?
      v. Copayment agreement necessary?

2. Read Statement of company standard terms and conditions

   a. Inform the bidder that:

      i. It will be bound by the desired delivery and payment schedules,

      ii. All other terms and conditions of the standard purchase order are required,

      iii. No variations from these terms in conditions will be accepted, and

      iv. The Company considers the agreement to be final as of this date, and is relying upon it.

3. Have an area at the end of the form for company personnel to initial; confirming that the statement had been read over the telephone to the specific individual indicated on the form. (Note that the failure to initial this area can later be used as proof that the statement was *not* read).

Figure 2.13 Sample Telephone Quote Checklist. *(continued on next page)*

By permission: The McGraw-Hill Company, New York, NY

Sample Telephone Quote Checklist
(Page 2 of 2)

4. Include a statement that the formal purchase order including all company's standard terms and conditions as well as special terms and conditions particular to the project will be prepared based upon the information herein and forwarded to the vendor for final execution.

5. Immediately fax and mail the completed and initialed form to the subvendor, and request immediate confirmation by signing and returning a copy of the form.

6. Note that generally, if the Telephone Quote form is not returned by the vendor within 10 days or if it is otherwise not responded to, the contract is likely to be considered unenforceable (Check with local laws) If, however, the vendor responds with modified terms, the contract may be confirmed to exist but may be subject to further negotiation.

7. If a vendor responds to the written confirmation of the Telephone Quote with either modifications to terms, additions to terms, or even its own Purchase Order form, it must be properly objected to in writing. The objection must include a remark that reaffirms the original statement that "no changes in terms will be allowed."

Figure 2.13 *(continued from previous page)* Sample Telephone Quote Checklist.
By permission: The McGraw-Hill Company, New York, NY

Section 3

Checklists for Residential Contractors

RESIDENTIAL CONSTRUCTION CHECKLISTS

Residential construction, by its nature, often requires slightly different inspection checklists. When building single-family houses, each building lot becomes an entirely new and different project. The buyer may have elected various options that change the footprint of the standard offering, or the exterior finish along with interior upgrades may significantly change its appearance from the house next door. Each site may present its own problems, drainage, soil conditions, and grade changes.

The reluctance of some residential contractors to switch from wood framing members to steel framing places more emphasis on checking finished framed walls for plumb, warped, or bowed studs, camber on wood floor and roof joists, and all the other problems associated with a product from nature.

Since the quality of framing has such a significant impact on the quality of the finished product, the Superintendent's Rough Trades Inspection Checklist may help to point out some of the items needing attention prior to the application of drywall or exterior wall treatment. Builders of residential units often have the added responsibility of inspecting the premises prior to owner acceptance or move-ins. The Apartment/Town House Inspection Checklist can serve not only as an inspection tool but also as a record of the condition of the unit prior to occupancy and a baseline for any punch list or warranty work down the road.

Although many of the Checklists in other sections of this book can be used for various construction component inspections, the following additional lists have more applicability for residential construction. Each of these inspection reports can also be placed in the permanent file of the unit to which it belongs thereby providing the necessary documentation for any future warranty or guarantee issues.

Residential Project Start-Up Checklist *(Page 1 of 2)*

Project ______________________

Location,Lot No.________________

Project Start date:______________

Project completion date:_________

_____ Contract signed

_____ Permits: City Bldg_____ Subdivision permit_______

Electrical Permit______ Plumbing permit _______

_____ Utilities required: Water_____ Sewer ______ Gas______ Electric______

_____ Layout: Lot:________ Bldg________

Building Layout

_____ Batter boards

_____ Lot stakes protected

_____ Utilities located ___ Water___Sewer____ Gas_____ Electric _____ Cable TV

Excavation

_____ Review cuts/fills per drawings

_____ Determine if borrow or removal of excess is required

_____ If structural borrow is required, located approved source

_____ Locate stockpile area that won't require re-locating

Concrete

_____ Mix design provided to supplier

_____ Verify finish on slabs,walks.garage slab

_____ Expansion joint material on hand and locations determined

_____ Reinforcing bars. Wire mesh as required on site

_____ Winter protection, tarps and heaters available if needed

_____ Verify any embeds, anchor bolts, conduits, pipe, etc

Figure 3.1 Residential Project Start-Up Checklist. *(continued on next page)*

Residential Project Start-Up Checklist *(Page 2 of 2)*

Prior to backfill

____ Determine where structural and non-structural fill is required

____ Foundation walls are properly cured

____ Penetrations thru walls are made and box-outs slushed in

____ Exterior of foundation walls are dampproofed as required

____ Underground utilities are pitched, stubbed, up as required

____ Gravel under slab is required depth and compacted

____ Wire mesh, if required, is supported up off gravel base

____ Expansion joints, perimeter insulation, as required, is in place

____ Vapor barrier of correct mil thickness is installed and adequately lapped

Vendors/Materials required when coming out of the Ground

____ Framing lumber

____ Plywood (grade), size,thickness, OSB

____ Trusses- ordered, engineering drawings approved

____ Windows

____ Doors, interior, exterior, garage, specialties

____ Insulation- wall- R value Ceiling- R value

____ Roofing type

____ Exterior finish–siding, masonry, shingles, EIFS-selection, thickness, type, color

Subcontractors

______Excavation

_____ Concrete

____ Electrical – power, telephone, communication

____ Plumbing & HVAC

____ Roofer

Figure 3.1 *(continued from previous page)* Residential Project Start-Up Checklist.

Apartment/Town House Inspection Checklist

Unit number:______________________________

Date: ______________________________

Entry-Foyer

___ Exterior door and hardware
___ Exterior door- all edges painted
___ Flooring and base
___ Light fixture
___ Walls, Ceiling (condition of drywall acceptable for finishes)

Living Room

___ Carpet and base
___ Walls (condition of drywall acceptable for finishes)
___ Ceiling (condition of drywall acceptable for finishes)
___ Paint and Caulk
___ Windows – open effortlessly, locks work
___ Sill - condition of paint, sanded properly, ends fitted
___ Screens – installed
___ Electrical switches/receptacles – flush to wall, sealed
___ Light fixtures - operative, secured, trim flush to ceiling
___ HVAC supply/returns – operative. flush to wall
___ Thermostat - secured to wall, functioning
___ TV/ Telephone/Data outlets installed, plates flush
___ Fire Protection – escutcheons tight to ceiling, protective cap off

General notes: ____

____

____

____

Kitchen

___ Vinyl flooring/base tight to wall
___ Walls (condition of drywall acceptable for finishes)
___ Ceiling (condition of drywall acceptable for finishes)
___ Pass-thru (if applicable) all sill edges smooth, drywall acceptable
___ Paint & caulk
___ Windows – open effortlessly, locks work
___ Screens installed
___ Light fixture installed, operational, lens crack-free
___ Electrical switches/receptacles- plates flush to wall
___ GFI outlet operative at wet areas
___ Base cabinets- doors, drawers operate smoothly
___ Wall cabinets – secured properly, doors operate smoothly
___ Counter Tops- laminate properly adhered, joints acceptable
___ Range installed, shelves in place, gas line wall penetration sealed
___ Range hood – secured, operative (fan and lights), filter installed
___ Garbage disposer- operative, sink plug
___ Refrigerator- level, shelves installed
___ Plumbing- Sink
___ Hot/cold water available
___ Flow acceptable, filter cleaned
___ Supply lines/valves–wall penetrations sealed, escutcheons on
___ Trap –no leaks observed
___ HVAC registers- operational, flush with wall
___ Fire protection- sprinkler heads, protective caps removed
___ Telephone/ TV/ data communication devices installed

Figure 3.2 Apartment/Town House Inspection Checklist. *(continued on next page)*

___ Exterior door and hardware – installed, functioning
___ Exterior door- all edges painted
___ Threshold- fits properly, sealed at edges, pitched away from house

General notes: ____
____
____
____

Master Bedroom

___ Carpet and base
___ Walls (condition of drywall acceptable for finishes)
___ Ceiling (condition of drywall acceptable for finishes)
___ Paint and caulking
___ Windows –open effortlessly, locks work
___ Screens installed
___ Sill- sanded properly, ends sealed, condition of paint
___ Electrical switches/receptacles- flush to wall, sealed
___ Light fixtures, operative, secured-all trim pieces flush to ceiling
___ HVAC- supply/returns- operative, grills flush to wall
___ Thermostat –secured to wall, functioning
___ TV/Telephone/data outlets installed, plates flush
___ Fire protection, escutcheons tight to ceiling, protective caps off
___ Closet – condition of drywall ceiling/walls acceptable for finishes
___ Shelf and poles in place
___ Flooring/base installed
___ Doors open and close properly
___ Interior corners- floor level meet properly
___ Light & switch/pull chain operational

General Notes: ____
____
____
____

Other Bedrooms

BR#2	BR#3	BR#4	
___	___	___	Carpet and base
___	___	___	Walls (condition of drywall acceptable for finishes)
___	___	___	Ceiling (condition of drywall acceptable for finishes)
___	___	___	Paint and caulking
___	___	___	Windows – open effortlessly, locks work
___	___	___	Screens installed
___	___	___	Electrical switches/receptacles – flush to wall, sealed
___	___	___	Light fixtures- operative, secured, trim installed
___	___	___	HVAC supply/returns – operative, flush to wall
___	___	___	TV/ Telephone/Data outlets installed, plates flush
___	___	___	Fire protection, escutcheon tight to ceiling, prot cap off
___	___	___	Closets- flooring and base
___	___	___	Closet walls, ceilings (drywall acceptable for finishes)
___	___	___	Closet paint
___	___	___	Doors, hardware open/close properly
___	___	___	Closet poles, shelves installed
___	___	___	Closet lights- switch/pull chain works, fixture secured

General Notes: ____
____
____
____

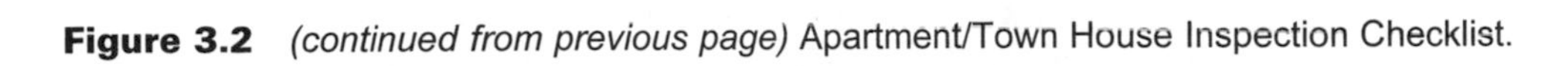

Figure 3.2 *(continued from previous page)* Apartment/Town House Inspection Checklist.

Bathrooms

Master Bath#2 Bath#3

Master	Bath#2	Bath#3	
___	___	___	Flooring and base - properly adhered, tight to wall
___	___	___	Walls and ceiling (drywall acceptable for finishes)
___	___	___	Door/hardware - edges painted, closes, locks
___	___	___	Vanity top and base (penetrations sealed)
___	___	___	Vanity light - operational, secured
___	___	___	Faucets- hot/cold flow, strainer clean
___	___	___	Medicine cabinet - secured, shelves installed
___	___	___	Tub-chip free, edges caulked, hot/cold, drain - O.K.
___	___	___	Shower - spout secured, diverter works
___	___	___	Ceramic tile - sealed at tub, check for damage
___	___	___	Grab bars installed, secured to wall
___	___	___	Toilet - operational, flushing works/stops
___	___	___	Toilet accessories, installed, secured
___	___	___	Exhaust fan operational
___	___	___	Electrical switches/receptacles – flush to wall, sealed
___	___	___	GFI receptacles installed - test
___	___	___	HVAC supply/returns - operative, flush to wall
___	___	___	Fire protection - escutcheons tight to ceiling, protect caps off

General Notes:____

____

____

____

Laundry Area

___ Hot/Cold water, drain connections installed
___ Overflow pan installed and piped
___ Doors/hardware installed - open/close properly
___ Special electrical receptacle installed, properly sealed
___ Light fixture installed, operational

General Notes: ____

____

____

____

Basement (if applicable)

___ Stairs- sturdy treads, handrail properly installed, sanded
___ Light fixtures installed, operational
___ No excessive slab or wall cracking
___ Basement free from water/moisture
___ No evidence of mold/mildew
___ All utility lines, ducts properly secured to walls, ceiling
___ Architect/engineer sign off on furnace/ air conditioning systems
___ Area broom clean, all debris removed

General Notes: ____

____

____

____

Figure 3.2 *(continued from previous page)* Apartment/Town House Inspection Checklist.

Concrete Checklist for Residential Building

Footings:

1. Ensure proper sub-grade elevation. __________
2. Confirm footing size and location in relation to the building layout. __________
3. Check forms for correct elevation, width and depth. Tape all measurements for length and width of building. After step down footings, landings, and slope footings are formed, set up builder's level to double check elevations. __________
4. Make sure forms are well braced, tight, free of debris and coated with proper release form material. __________
5. Check size and placement of reinforcing steel. Make sure rebars are correctly fabricated and installed and will remain up off bottom of excavate when poured. ________
6. Schedule necessary inspections. ________

Prior to Pour:

1. Check for proper thickness, correct steel, specified mix design of ready mix concrete. __________
2. Check anchor bolts, hold-downs, post anchors, column bases for proper installation – location and embedment. __________
3. Make sure all depths are correct per plans and soils report. __________
4. Make sure trenches are free of dirt and debris. __________
5. Check for any eased edges at stem walls, step downs. __________
6. Check installation of any expansion joints, joints at split levels, steps, grade breaks. __________
7. Be prepared to take slump test when concrete is delivered and to take test cylinders. __________
8. Get ticket for each batch of concrete delivered. __________
9. Do not allow water to be added to the transit mix. __________

Figure 3.3 Concrete Checklist for Residential Building.

Superintendent's Rough Trades Inspection Checklist

❑ Check here when rough heat inspection is complete

❑ Check here when rough electrical inspection is complete

❑ Check here when rough plumbing inspection is complete

Framing Inspection

❑ Check wall lines at floor and ceiling

❑ Check window and door opening sizes- width___ height___

❑ Check for special shear blocking

❑ Check joist blocking, spacing, size and direction

❑ Check all special hardware or framing details

❑ Check nailing schedules

❑ Check all metal connections

❑ Check plumb at all corners

❑ Check closet door openings sizes – width ___ height ___

❑ Check stairs for stringer size, clearance for skirt boards and drywall. Check for proper stair stringer nailing and blocking. Also check tread depth, width and riser height

❑ Check medicine cabinet height, width, location

❑ Check tub and shower backing. Check flat block backing for shower and tub jambs

❑ Check medicine cabinet blocking

❑ Check towel bar, toilet paper holder, towel rings, toothbrush holder, soap dish - location and blocking

❑ Check for Handicap provisions - grab bar blocking

❑ Check all dropped ceilings for correct height___ width___ depth ___ and proper joist size and grade

❑ Check all post to beam connections

❑ Make sure all window headers are furred level for drywall. Make sure the margins are even

❑ Make sure windows are set square and even to openings

❑ Check all ceilings with an attic above to make sure they have the following:
A- Attic access
B- Strong backs.

Figure 3.4 Superintendent's Rough Trades Inspections Checklist. *(continued on next page)*

Superintendents' Rough Trades Inspection Checklist Page 2

- ❑ Check for all lath or exterior siding backing
- ❑ Check fireplace faces for drywall backing
- ❑ Check all handrail backing (flat blocks)
- ❑ Verify that all plumbing and heating straps align, are flush with wall and will not cause drywall to bulge
- ❑ If there is a pair of windows or a door and a window side by side, verify that headers line up. Align properly inside and outside
- ❑ Verify that there are no extra anchor bolts protruding up in door ways or outside under patio door sills or mudsills
- ❑ Verify that water service risers are plumb and will fit inside wall (if applicable)
- ❑ Verify that all heating ducts will fit inside wall
- ❑ Verify that all staples or other fastening devices that held temporary braces are removed from exterior wood door jambs
- ❑ Check sizes of all roof gable vents, foundation vents, storage and utility room vents
- ❑ Verify that exterior wood frames are back primed
- ❑ Verify that garage door jambs are fastened securely and grade of lumber used is correct
- ❑ Verify that all window and door frames are properly set and flashed
- ❑ Verify that any exterior columns are plumb and have proper size trim applications
- ❑ Verify that all shower door openings are plumb
- ❑ Nail all squeaky floors. Walk and mark areas that need correction (delaminated plywood, too tight joints, excessive joints)
- ❑ Verify elevations for pop-outs around windows and doors
- ❑ Verify elevations for correct exterior wall treatment (siding, stucco, etc)
- ❑ Check nailing of all king studs to headers
- ❑ Check nailing of corners, channels and intersecting walls
- ❑ Check fire blocking at dropped ceilings, soffits and stairs
- ❑ Check exterior wall for straightness and plumb
- ❑ Check starter board above entries and overhangs for damaged wood,shinners, knotholes and proper blocking
- ❑ Check for damaged rafter tails

Figure 3.4 *(continued from previous page)* Superintendent's Rough Trades Inspections Checklist.

Superintendent's Rough Trades Inspection Checklist Page 3

❑ Check for damaged bird stops, pressure blocks, barge or fascia block and missing blocks

Other Comments/Observations:

____

____

____

____

____

____

____

____

____

____

____

____

____

____

____

____

____

____

____

____

Lot or House Number:____________________

Date of Inspection: ____________________

Inspected by: ____________________

Figure 3.4 *(continued from previous page)* Superintendent's Rough Trades Inspections Checklist.

Floor Covering Checklist for Residential Builders

When laying entry vinyl, ceramic tile, or wood parquet flooring

- ❑ Read contract to confirm materials and installation procedures
- ❑ Be sure flooring material is laid square with walls and openings. Make sure it has an even margin at walls
- ❑ Make sure step downs have finished edges
- ❑ Confirm the correct grout color, floor finish or pattern
- ❑ Check for defects or discolorations in grout, floor finish or pattern
- ❑ Protect flooring with cardboard, building paper or polyethylene as soon as possible after installation
- ❑ Re-install baseboard or base shoe as required
- ❑ Inspect flooring surface for damage, replace as required

When installing resilient flooring

- ❑ Check floor plan, flooring contract or homeowner selections for correct pattern, color
- ❑ Ascertain that material is cut tight to walls with no obvious seams
- ❑ Make sure vinyl flooring is caulked at tubs, showers, water closet
- ❑ Check for cuts around kitchen appliances if appliances have been installed prior to flooring installation . Preferable method is to install flooring first.
- ❑ Roll vinyl to eliminate air bubbles
- ❑ Install transition strips where flooring abuts dissimilar flooring materials
- ❑ Check manufacturer's specifications for proper seam sealer
- ❑ Insure that flooring has been installed with no resulting damage; replace as required
- ❑ Observe that cleaning and waxing has been satisfactorily completed

When installing carpet

- ❑ Check with flooring contractor to verify type, style, manufacturer, quality, weight, and color of carpet
- ❑ Insure that subfloor is securely fastened to structure; there is no delamination of plywood sheets; plywood is installed with acceptable expansion spacing
- ❑ Insure that tack strip is properly secured to floor
- ❑ Check for correct type of pad or adhesive, if direct application
- ❑ Insure that seams are correct and trimmed
- ❑ Ensure that carpet is properly stretched
- ❑ Are transition strips, as required, installed securely?
- ❑ Insure that carpet installer did not damage adjacent surfaces
- ❑ Check to ensure that all door undercuts are acceptable. If additional undercuts are required, they are performed by an experienced carpenter to avoid damage to doors
- ❑ Carpet is vacuumed and inspected for damage,

Figure 3.5 Floor Covering Checklist for Residential Builders.

Town House - Exterior Checklist

Front/Rear Elevations:

___ Door frame-secured to structure, painted and caulked
___ Frame isolated from sill and caulked
___ If transom, check that glass is secured, frame caulked
___ Steps (if applicable) are structurally sound, rail per code
___ Window frames caulked, sills structurally sound

Exterior wall (if masonry)

___ Mortar joints sound
___ No evidence of effluoresence
___ All penetrations are caulked, lights, hose bibs, etc

Exterior wall (if vinyl siding)

___ Check that starter strip is level
___ Siding secured to substrate with nails partially driven home
___ ¼" clearance around all openings
___ Do not caulk overlap joints
___ Do not caulk where panels meet receivers
___ No stapling or fastening thru face of siding
___ Flashings around windows extend beyond nail flanges
___ J- channel installed around windows to accept siding
___ Gable J channel trim to receive siding

Front/ Rear Concrete Walkways – Rear Yard Areas

___ Expansion joints in walks where required
___ Concrete surfaces- broom finish. No evidence of spalling
___ No excessive cracking in concrete
___ Walk pitches away from house, or if grade does not allow, pitches to one side
___ Landscaped area graded to allow water to flow away from structure
___ Downspouts connected to underground drain or to splash block
___ If sodded, strips are rolled in not merely placed on surface.
___ Sod placed on slopes to be staked
___ Yard trees to be staked and guyed

Figure 3.6 Town House - Exterior Checklist.

Section 4

Subcontractor Interview Forms

The Subcontractor Interview Form

All too often, we interview and negotiate with subcontractors on the basis of "plans and specifications". But as we all know, the plans and specs may not contain all of the work that is required of a subcontractor. The plans and specs may contain some omissions that need to be filled in to obtain the full scope of work and avoid those potential change orders as work gets underway, when the subcontractor says, "Well, the plans and specs didn't call for that and I didn't include that work in my price." Do you recall a time when the concrete subcontractor failed to look at the mechanical and electrical drawings and didn't find the cast-in-place equipment pads or the concrete transformer pad? Instead of an "extra", this could have been included in their contract scope, often at little or not increase in price, if this work had been discussed at an interview.

By reviewing the plans and specifications before meeting with a subcontractor to review their proposal, it is a good idea to jot down some of the essential items you need to have them include. During this review, you may also come across some gaps in the particular scope of work so that you can include additional work in that subcontractor's proposal- hopefully at no increase in price. One such item that comes to mind is "starters" – the electrical controllers required to start a motor. The HVAC specification section will undoubtedly include some motors for air-handling units or condensing units, but may be silent on which subcontractor furnishes those items. And the electrical specification may state that the electrician is to furnish starters for all equipment to be supplied in their spec section but nothing else. So unless the subject is specifically mentioned during an interview, the HVAC sub may claim that this is an electrical item and the electrical subcontractor could deny responsibility since they did not supply the HVAC equipment. By using one of these interview forms, you may be able to settle matters such as this and reduce the potential for "extras."

The need to prepare for interviews with subcontractors is essential if you want to insure that you have adequately compared one sub's proposal with another or made the necessary adjustments to do so. Also note that you have included any additional work, and that you arrived at the most competitive bid.

The Subcontractor Interview Forms in this section of the book will be of some assistance in this process. Of course, some customizing will be necessary to fit the project at hand, but each form will provide you with some basic elements that should help you along. There is also a blank form that can be used to create your own customized version.

SUBCONTRACTOR NEGOTIATION FORM

PROJECT		
TRADE **Acoustical Treatment**	SPECIFIED SECTIONS	DATE
SUBCONTRACTOR	REPRESENTED BY	(AREA CODE) TELEPHONE NO.
BASE BID AMOUNT	ADDENDUM NO.	
ALTERNATES (1)	UNIT PRICES (1)	UNIT PRICES (6)
(2)	(2)	(7)
(3)	(3)	(8)
(4)	(4)	(9)
(5)	(5)	(10)
SALES TAX	INSURANCE	

SCOPE OF WORK Including but not limited to the following:

This form must be completely filled out and signed by the subcontractor and general contractor's representative to provide a record of this negotiation meeting.

ITEM	YES	NO	EXPLANATION AND/OR COMMENTS
1. Includes all necessary labor, materials, and equipment to complete all acoustical tile work in accordance with the contract documents.			
2. Subcontractor has reviewed specification section ________ and the general and special conditions and agrees to all terms relating to their trade.			
3. Insurance certificates as required by the contract specifications will be submitted.			
4. All hoisting and scaffolding required for the subcontractor's work will be provided by the subcontractor.			
5. The subcontractor has received and reviewed the construction schedule and accepts that portion relating to their trade.			
6. The subcontractor has received, read, and accepts the general contractor's safety program.			
7. All layout to be provided by subcontractor.			
8. Alternates as attached are accepted. Unit prices, as required, are attached.			
9. Rubbish removal to a dumpster, at location directed by the general contractor, to be performed on a daily basis			

	FINAL AGREED AMOUNT
GENERAL CONTRACTOR'S REPRESENTATIVE SIGNATURE	SUBCONTRACTOR'S REPRESENTATIVE SIGNATURE

ORIGINAL

Figure 4.1 Subcontractor Form — Acoustical Treatment. *(continued on next page)*

CONTINUATION SHEET

SUBCONTRACTOR NEGOTIATION FORM

Page 2 of 2

ITEM	YES	NO	EXPLANATION AND/OR COMMENTS
10. Subcontractor to coordinate their work with related trades - all discrepancies to be reported to general contractor when they occur.			
11. Provide mock-up, if required.			
12. Provide any black iron supports required for this subcontractor's work.			
13. Provide identification markers approved by general contractor for all accessible panels.			
14. All ACT soffit work is included in this contract.			
15. Subcontractor to provide all lay-in batt insulation required above acoustical ceiling work.			
16. Supply all window pockets for drapes or blinds as indicated on the contract drawings.			
17. Remove and replace any tiles required by other trades working above the ACT ceilings.			
18. All tiles in an area to be of same lot number.			
19. Furnish and install any wood nailers required for ACT work.			
20. Provide tenting for lay-in electrical fixtures as required.			
21. Provide sufficient attic stock for each type of tile as indicated in the specifications or in reasonable amounts as directed by the general contractor.			
22. Subcontractor is to return the subcontract agreement, if awarded, within ____ days. No work is to commence until a signed contract has been received and accepted by the general contractor.			

Figure 4.1 *(continued from previous page)* Subcontractor Form — Acoustical Treatment.

SUBCONTRACTOR NEGOTIATION FORM

PROJECT		
TRADE	SPECIFIED SECTIONS	DATE
SUBCONTRACTOR	REPRESENTED BY	(AREA CODE) TELEPHONE NO.
BASE BID AMOUNT	ADDENDUM NO.	
ALTERNATES (1)	UNIT PRICES (1)	UNIT PRICES (6)
(2)	(2)	(7)
(3)	(3)	(8)
(4)	(4)	(9)
(5)	(5)	(10)
SALES TAX	INSURANCE	

SCOPE OF WORK Including but not limited to the following:

This form must be completely filled out and signed by the subcontractor and the general contractor's representative to provide a record of this negotiation meeting.

ITEM	YES	NO	EXPLANATION AND/OR COMMENTS
1.			
2.			
3.			
4.			
5.			
6.			
7.			
8.			
9.			

	FINAL AGREED AMOUNT
GENERAL CONTRACTOR"S REPRESENTATIVE SIGNATURE	SUBCONTRACTOR,S REPRESENTATIVE SIGNATURE

ORIGINAL

Figure 4.2 Blank Subcontractor Negotiation Form for Other Trade Applications. *(continued on next page)*

CONTINUATION SHEET

SUBCONTRACTOR NEGOTIATION FORM

Page 2 of 2

ITEM	YES	NO	EXPLANATION AND/OR COMMENTS
10.			
11.			
12.			
13.			
14.			
15.			
16.			
17.			
18.			
19.			
20.			
21.			
22.			

Figure 4.2 *(continued from previous page)* Blank Subcontractor Negotiation Form for Other Trade Applications..

SUBCONTRACTOR NEGOTIATION FORM Page 1 of 2

PROJECT		
TRADE Cabinet Work	SPECIFIED SECTIONS	DATE
SUBCONTRACTOR	REPRESENTED BY	(AREA CODE) TELEPHONE NO.
BASE BID AMOUNT	ADDENDUM NO.	
ALTERNATES (1)	UNIT PRICES (1)	UNIT PRICES (6)
(2)	(2)	(7)
(3)	(3)	(8)
(4)	(4)	(9)
(5)	(5)	(10)
SALES TAX	INSURANCE	

SCOPE OF WORK Including but not limited to the following:

This form must be completely filled out and signed by the subcontractor and the general contractor's representative to provide a record of this negotiation meeting.

ITEM	YES	NO	EXPLANATION AND/OR COMMENTS
1. Include all necessary labor, materials, equipment, and tools as required to furnish and install finished cabinet work in accordance with the contract documents.			
2. Comply with lumber and plywood standards as specified.			
3. Submit all factory markings, certificates, and samples as required.			
4. Pressure treatment and fire retardant treatment as specified or required.			
5. Furnish and install all specified fastenings and anchorage material.			
6. Furnish and install all radii as per plans and specifications.			
7. All furnished work damaged, marred, chipped, scratched, or damaged by this subcontractor or others will be wholly replaced by this subcontractor at no cost to the general contractor or owner.			
8. Furnish and install all finished woodwork and plastic laminate.			
9. Furnish and install cabinet hardware			

	FINAL AGREED AMOUNT
GENERAL CONTRACTOR'S REPRESENTATIVE SIGNATURE	SUBCONTRACTOR'S REPRESENTATIVE SIGNATURE

ORIGINAL

Figure 4.3 Subcontractor Form — Cabinet Work. *(continued on next page)*

CONTINUATION SHEET

SUBCONTRACTOR NEGOTIATION FORM Page 2 of 2

ITEM	YES	NO	EXPLANATION AND/OR COMMENTS
10. Install finish hardware. General contractor and subcontractor to agree on scope of work and price.			
11. Furnish and install all brackets and supports, metal, or wood required for millwork supplied by this subcontractor.			
12. All hoisting and/or labor to distribute/store the items being installed by this subcontractor are the responsibility of this subcontractor.			
13. Protection of all installed millwork and all doors (if applicable) remain the responsibility of this subcontractor.			
14. Hourly rates for any extra or T&M work, including burden and _____ % overhead and profit will be in effect during the life of this contract.			
15. All work to be in accordance with all local, state, and federal laws and ordinances.			
16. All scaffolding required for the work covered by this agreement is included in the subcontractor's price.			
17. Layout and engineering are the responsibility of this subcontractor.			
18. Taxes are included, or in the case of a not-for-profit owner, taxes are excluded.			
19. Additional percentage for extra work is as stated here.			
20. Working hours for this project are as follows:			
21. Retainage percentage will be ____%.			
22. Subcontractor verifies that they have included all labor and material escalation charges to cover the scope of their work.			
23. Subcontractor acknowledges and accepts the general contractor's and OSHA safety regulation and programs and will abide by all EEO rules and regulations. By signing this interview for, the subcontractor agrees to abide by its terms and conditions which will be incorporated in the subcontract agreement, a copy of which is attached.			

Figure 4.3 *(continued from previous page)* Subcontractor Form — Cabinet Work.

SUBCONTRACTOR NEGOTIATION FORM

PROJECT		
TRADE **Caulking & Sealants**	SPECIFIED SECTIONS	DATE
SUBCONTRACTOR	REPRESENTED BY	(AREA CODE) TELEPHONE NO.
BASE BID AMOUNT	ADDENDUM NO.	
ALTERNATES (1)	UNIT PRICES (1)	UNIT PRICES (6)
(2)	(2)	(7)
(3)	(3)	(8)
(4)	(4)	(9)
(5)	(5)	(10)
SALES TAX	INSURANCE	

SCOPE OF WORK Including but not limited to the following:

This form must be completely filled out and signed by the subcontractor and the general contractor's representative to provide a record of this negotiation meeting.

ITEM	YES	NO	EXPLANATION AND/OR COMMENTS
1. Include all necessary labor, materials, and equipment to complete all caulking and sealant work in accordance with the contract documents.			
2. Subcontractor has reviewed specification section _________ and the general and special conditions and agrees to all terms relating to their trade.			
3. Insurance certificates per contract specification will be provided prior to commencement of work.			
4. All hoisting and scaffolding required for the subcontractor's trade will be provided by the subcontractor.			
5. Subcontractor has reviewed the company's subcontract agreement and agrees to its terms.			
6. Subcontractor has received, reviewed the construction schedule, and accepts that portion pertaining to their trade.			
7. Subcontractor has received, read, and accepts the general contractor's safety program.			
8. Until such time as general contractor can provide sufficient lighting, all such requirements are the responsibility of this subcontractor.			
9. Alternates are attached and accepted.			

	FINAL AGREED AMOUNT
GENERAL CONTRACTOR'S REPRESENTATIVE SIGNATURE	SUBCONTRACTOR'S REPRESENTATIVE SIGNATURE

ORIGINAL

Figure 4.4 Subcontractor Form — Caulking and Sealants. *(continued on next page)*

CONTINUATION SHEET

SUBCONTRACTOR NEGOTIATION FORM Page 2 of 2

ITEM	YES	NO	EXPLANATION AND/OR COMMENTS
10. Rubbish removal to take place daily and deposited in a dumpster as designated by the general contractor.			
11. Unit prices, if applicable, have been submitted and are attached.			
12. Subcontractor to provide submittal schedule to include samples, product data.			
13. Storage of all combustible materials to be in Fire Department-approved containers.			
14. Additional work percentages:___% profit, ____ % overhead			
15. Work to include, but not limited to, window frames, exterior door frames, duct penetrations, lintels, louvers, control joints, fascias, and flashings not caulked by roofer.			
16. Work to include, but not limited to, interior expansion joints, control joints, construction joints, perimeter joints.			
17. Work to include, but not limited to, exterior concrete and masonry expansion and control joints.			
18. Masking of adjoining surfaces, as required, and removal of same.			
19. Cleaning of all open joints to be caulked as required.			
20. Caulking of all precast concrete components.			
21. Caulking all of vanity tops and backsplashes in bathrooms and kitchens.			
22. Subcontractor to return subcontract agreement within ___ days of receipt. No work can commence until a signed contract has been received and approved by the general contractor.			

Figure 4.4 *(continued from previous page)* Subcontractor Form — Caulking and Sealants.

SUBCONTRACTOR NEGOTIATION FORM

PROJECT		
TRADE **Concrete, Cast-in-place**	SPECIFIED SECTIONS	DATE
SUBCONTRACTOR	REPRESENTED BY	(AREA CODE) TELEPHONE NO.
BASE BID AMOUNT	ADDENDUM NO.	
ALTERNATES (1)	UNIT PRICES (1)	UNIT PRICES (6)
(2)	(2)	(7)
(3)	(3)	(8)
(4)	(4)	(9)
(5)	(5)	(10)
SALES TAX	INSURANCE	

SCOPE OF WORK Including but not limited to the following:

This form must be completely filled out and signed by the subcontractor and general contractor's representative to provide a record of this negotiation meeting.

ITEM	YES	NO	EXPLANATION AND/OR COMMENTS
1. Quantity check: Total Concrete:_______________ cy Foundation concrete: _________ cy Slab on grade concrete:________ cy Slab(s) on deck concrete:________cy Misc. concrete:_______________ cy			
2. Quantity check: Site concrete:_______________cy Mech/Electrical concrete_______ cy			
3. Reinforcing steel: Rebar tonnage:__________ lbs Welded Wire Mesh_______ sf Caps on all vertical bars, accessories, chairs, etc.			
4. Concrete mix design per specifications: Stone Lightweight			
5. Admixtures per specifications			
6. Special aggregates, cement color			
7. Slump as specified- no exceptions taken			
8. Forms - specify type for structural concrete. Liners for architectural concrete			
9. Chamfer strips and rustications			

	FINAL AGREED AMOUNT
GENERAL CONTRACTOR'S REPRESENTATIVE SIGNATURE	SUBCONTRACTOR'S REPRESENTATIVE SIGNATURE

ORIGINAL

Figure 4.5 Subcontractor Form — Concrete, Cast-in-Place. *(continued on next page)*

CONTINUATION SHEET

SUBCONTRACTOR NEGOTIATION FORM Page 2 of 2

ITEM	YES	NO	EXPLANATION AND/OR COMMENTS
10. Scaffolding, pumps, hoisting and lifting as required			
11. Provide layout and install box-outs, sleeves (by others), framed openings, embeds (by others) as required.			
12. Mock-ups as required			
13. Dowels for other trades			
14. Shoring-installation and removal only when written approval is received from engineer			
15. Additional work: ________% overhead, _______ % profit			
16. Schedule of values to be submitted for approval by general contractor			
17. Subcontractor to provide temporary heat and winter protection for their work, as required.			
18. Subcontractor to provide line, grade, and all engineering required for the placement of their work.			
19. Contractor to coordinate and receive approval of placement of field trailer and lay-down areas for their work.			
20. All debris, excess materials to be removed from site.			
21. Subcontractor has received, read, and accepts subcontract agreement. Also he has received, read, and accepts all specification provisions including General and Special Conditions.			
22. Subcontractor to sign subcontract agreement, if awarded within ____ days of receipt, and cannot commence work on site until a fully executed contract is received by them.			
Detail sheets, when applicable are attached hereto and become a part of this negotiation session when initialed by subcontractor and (company).			

Figure 4.5 *(continued from previous page)* Subcontractor Form — Concrete, Cast-in-Place.

Concrete, Cast-in-Place Detail Sheet No. 1

Foundation Work

- ❑ Hand work for final grade
- ❑ Pits as required
- ❑ Sumps as required
- ❑ Mud mats or rat slabs as required
- ❑ Set column anchor bolts
- ❑ Set leveling plates
- ❑ Set beam anchor bolts and bearing plates
- ❑ Dewater as required for your work
- ❑ Form beam pockets as required
- ❑ Box out for columns
- ❑ Provide dowels for attachment by other trades
- ❑ Install embedded items supplied by others

Slab on Grade

- ❑ Fine Grade as Required
- ❑ Compact as required after installation of underslab utilities
- ❑ Insure that all mechanical and electrical stub-ups are above grade and are plumb
- ❑ Verify that correct depth of porous compacted fill has been placed
- ❑ All anchor bolts, embedded items are properly and accurately installed
- ❑ Perimeter insulation installed
- ❑ Provide all box-outs as required
- ❑ Rebars and WWM supported properly and at elevation specified
- ❑ Vapor barrier of correct mil thickness, installed, lapped and protected against puncture
- ❑ Finishing is performed per specification, steel trowel finish and to level tolerances of ___ inch per 10 feet in all directions
- ❑ Saw cuts, if Zip Strips not used, are performed per specifications

Slab-on-Metal Deck

- ❑ Verify that edge forms and pour stops are installed as required
- ❑ Verify that all box-outs, sleeves, inserts are installed
- ❑ Verify that all openings over 144 square inches are framed. Inspect with GC
- ❑ All embedded items furnished to this subcontractor for installation
- ❑ Shear studs, if composite deck, are in place
- ❑ Finishing by steel trowel to meet tolerance of _____ inch per 10 feet in each direction

Miscellaneous Building Concrete Items

- ❑ Furnish and install all inserts, dovetail slots and reglets for other trades as required
- ❑ Grout column billets
- ❑ Close column pockets
- ❑ Close beam pockets
- ❑ Fire stopping, as required
- ❑ Form and grout elevator saddle recesses
- ❑ Clean stair pans prior to concreting
- ❑ Set non-slip nosings
- ❑ Apply non-slip dust or shakes as required

Figure 4.6 Concrete, Cast-in-Place Detail Sheet No. 1.

Concrete, Cast-in-Place Detail Sheet No. 2

Sitework

- ❏ Walks adjacent to building
- ❏ Site sidewalks
- ❏ Curbs
- ❏ Retaining Walls
- ❏ Site stairs
- ❏ Flagpole bases
- ❏ Sign bases
- ❏ Slabs as sub-base for stone paving
- ❏ Entrance canopy slabs
- ❏ Benches
- ❏ Headwalls
- ❏ Electrical and Mechanical equipment pads
- ❏ Concrete thrust blocks
- ❏ Aprons for asphalt paving
- ❏ Electrical and Mechanical pits, pull boxes
- ❏ Site light bases
- ❏ Pipe bollards
- ❏ Planters

Subcontractor- Date ______________________________

General Contractor- Date ___________________________

Figure 4.7 Concrete, Cast-in-Place Detail Sheet No. 2.

SUBCONTRACTOR NEGOTIATION FORM

Page 1 of 2

PROJECT

TRADE	SPECIFIED SECTIONS	DATE
Electrical		
SUBCONTRACTOR	REPRESENTED BY	(AREA CODE) TELEPHONE NO.
BASE BID AMOUNT	ADDENDUM NO.	
ALTERNATES	UNIT PRICES	UNIT PRICES
(1)	(1)	(6)
(2)	(2)	(7)
(3)	(3)	(8)
(4)	(4)	(9)
(5)	(5)	(10)
SALES TAX	INSURANCE	

SCOPE OF WORK Including but not limited to the following:

This form must be filled out completely and signed by subcontractor and the general contractor's representative to record this negotiation meeting.

ITEM	YES	NO	EXPLANATION AND/OR COMMENTS
1. Include all necessary labor, materials, and equipment, tools, and appurtenances as required to complete the electrical work in accordance with the contract documents.			
2. All systems include secondary distribution systems light and power, emergency generator, voice/data communication, fire alarm, and connections to all equipment as required.			
3. Main distribution panel, panelboards, safety switches, motor control centers, motor starters, switches, and contactors.			
4. Wiring for elevators, distribution feeders, fused disconnects, disconnect switch interlocks.			
5. Lighting fixtures-furnish and install building and site with lamps. Anchor bolts for site lights to be supplied for installation by others.			
6. Communication systems including all required conduit.			
7. Complete fire alarm system including smoke alarms required for building and mechanical systems.			
8. Smoke/fire damper monitoring systems.			
9. Security system Sound system Closed circuit TV system			

	FINAL AGREED AMOUNT
GENERAL CONTRACTOR'S REPRESENTATIVE SIGNATURE	SUBCONTRACTOR'S REPRESENTATIVE SIGNATURE

ORIGINAL

Figure 4.8 Subcontractor Form — Electrical. *(continued on next page)*

CONTINUATION SHEET

SUBCONTRACTOR NEGOTIATION FORM Page 2 of 2

ITEM	YES	NO	EXPLANATION AND/OR COMMENTS
10. Electrical heating- furnish and install U.O.N.			
11. Switchboard and grounding system.			
12. Motor control & control wiring- coordinate with GC, mechanical trades, and engineer.			
13. Temperature control wiring - coordinate with mechanical trades and engineer.			
14. Power wiring for sprinkler system.			
15. Emergency power system to include day tank.			
16. Pipe heat tracing system.			
17. Unloading, hoisting and scaffolding for all work.			
18. Kitchen equipment wiring to include final connections to all equipment including hoods.			
19. Site conduits, cabling, wiring, concrete encasements as required.			
20. Rubbish removal to occur daily with debris deposited in dumpster provide by GC.			
21. Additional work: ____ % overhead,_____% profit			
22. Subcontract has received, read, and accepts GC's subcontract agreement and safety program.			
23. Provide all O&M manuals, parts lists, attic stock, and spare parts as required. 24. Provide and maintain temporary light and power. Sufficient lumens for trades as required. Remove when permanent power is obtained.			

Figure 4.8 *(continued from previous page)* Subcontractor Form — Electrical.

SUBCONTRACTOR NEGOTIATION FORM

PROJECT		
TRADE **Elevator**	SPECIFIED SECTIONS	DATE
SUBCONTRACTOR	REPRESENTED BY	(AREA CODE) TELEPHONE NO.
BASE BID AMOUNT	ADDENDUM NO.	
ALTERNATES (1)	UNIT PRICES (1)	UNIT PRICES (6)
(2)	(2)	(7)
(3)	(3)	(8)
(4)	(4)	(9)
(5)	(5)	(10)
SALES TAX	INSURANCE	

SCOPE OF WORK Including but not limited to the following:

This form must be completely filled out and signed by the subcontractor and general contractor's representative to record the negotiation meeting.

ITEM	YES	NO	EXPLANATION AND/OR COMMENTS
1. Subcontractor has received complete set of drawings and specifications including general and/or special conditions and assumes responsibility for all respective work.			
2. Design Features: Passenger Freight Hospital Special			
3. Elevator Type: Hydraulic Traction Geared/Gearless Holeless			
4. Design loads (lbs)			
5. Design Speed(s)- fpm			
6. Travel distance			
7. Number of stops			
8. Machine location (see attached checklist for Machine Room)			
9. Platform size(s) as specified			
10. Car and hoistway sizes, types, material, power operation, as specified and acceptable for installation			
11. Pit as specified with sump, ladder, electrical power, dimensions, and depth to meet Code			

	FINAL AGREED AMOUNT
GENERAL CONTRACTOR'S REPRESENTATIVE SIGNATURE	SUBCONTRACTOR'S REPRESENTATIVE SIGNATURE

ORIGINAL

Figure 4.9 Subcontractor Form — Elevator. *(continued on next page)*

CONTINUATION SHEET

SUBCONTRACTOR NEGOTIATION FORM

Page 2 of 2

ITEM	YES	NO	EXPLANATION AND/OR COMMENTS
12.Hoisting and car equipment as specified: Rails/ cables Sheaves and supporting members Counterweights Governor and safety devices Automatic Terminal Stopping device Pit stop switch Recabling hitches			
13. Machine room equipment as specified: Type of machine Hoisting motor Motor controls Motor generator set Controller master selector Stop counter, car mileage indicator Tool board Hydraulic pump(s)			
14. Operating systems as specified: Selective Collective Duplex Selective Collective Attendant Operation Emergency elevator operation Top of car operating device Emergency lighting			
15. Signal systems as specified: Car position indicator Alarm bell Hall lanterns Starter panels Emergency Terminal Return Control Hoistway access switches			
16. Communication systems as specified: Intercommunication system Telephone (dedicated line by owner)			
17. Car equipment as specified: Car frame and auxiliary supports Platform- size and type Toe guard Sill Finish floor covering Door finish Walls and base Ceiling			
18. Other conveying types Escalators Dumbwaiters Moving ramps and people movers Parcel lifts			
19. Subcontractor to provide schedule of all submittals and projected delivery dates after receipt of approvals			
20. Subcontractor has received, read, and accepts the general contractor's safety program			
21. Subcontractor to furnish a detailed Schedule of Values for approval			
22. Subcontractor is to return signed copies of the subcontract agreement within____ days of receipt. No on-site work can commence until a fully executed contract is received.			

Figure 4.9 *(continued from previous page)* Subcontractor Form — Elevator.

Elevator Hoistway Checklist

❑ Hoistways serving more than three (3) floors shall be provided with means for venting smoke and hot gases directly to the outside or through non-combustible ducts to the outer air in case of fire.

❑ Vents shall be located in the side of the hoistway directly below the roof or directly over the hoistway.

❑ The clear area of the vents shall be not less than three and one-half (3 ½) percent of the area of the hoistway and in no case less than three (3) square feet for each elevator car, whichever is greater. Note: More stringent building codes will take precedence over elevator code requirements.

❑ Special attention is required for hoistway ventilation exposed to the weather; extra precaution must be taken to prevent entry of blowing rain.

❑ Vents shall have insect screens to prevent entry of insect, birds, etc.

❑ Hoistway enclosures shall have substantially flush surfaces on the hoistway side subject to the following:

1. On sides not used for loading and unloading
   - Recesses except those required for installation of elevator equipment
   - Beams, floor slabs or other building construction making an angle less than 75 degrees with the horizontal shall not project more than 2 inches (51mm) inside the hoistway enclosure unless the top surface of the projection is beveled at an angle less than 75 degrees with the horizontal. The top surfaces of intersections created by diagonal bracing, if projected more than 2 inches (51mm), shall be beveled at an angle not less than 75 degrees with the horizontal.

❑ Drywall construction, if used, shall be non-combustible construction only. Responsibility for maintaining the fire rating for any drywall penetrations is the responsibility of the drywall installer. Support for rail brackets for floors greater than 14'-0" and at the top of the hoistway is to be provided. More stringent local building codes will take precedence over the elevator code.

❑ Hoistway door entrance frames, headers and jambs shall be grouted solid to maintain fire rating of the hoistway.

❑ All voids, holes, slots, etc in the hoistway shall be grouted or pointed up to maintain fire rating of the hoistway.

❑ All nails, snap-ties, form straps and wood shall be removed from the hoistway.

❑ Where permitted, glass curtain walls may be used in the elevator hoistway provided the walls are of laminated glass (ANSI 297.1). Glass must extend a minimum of 10'-0" above each floor landing or as required by local codes.

❑ Only equipment required for the operation of the elevator is permitted inside the elevator hoistway.

❑ All landing floor and elevator lobbies shall be completed prior to any inspections.

❑ Proper hoistway dimensions must be maintained to ensure code-required refuge space when elevator is at its furthest travel in each direction.

Figure 4.10 Elevator Hoistway Checklist.

Elevator Machine Room Checklist

These forms are subject to adjustment for local and state regulations.

- ☐ Only machinery/equipment required for the operation of the elevator shall be permitted in the elevator machine room.
- ☐ Pipes or ducts conveying gases, vapors, or liquid or electrical equipment, wiring, piping, and controls not used for the operation of the elevator are prohibited in the elevator machine room.
- ☐ The elevator machine room shall be free of refuse and shall not be used for storage of any materials not related to the elevator equipment.
- ☐ Machine Room doors to have at least ½ hour fire rating. Doors to be self-closing and self-locking.
- ☐ Minimum machine door height is 7"-0". Minimum door width is 36". Door lock shall be of type that can be opened from inside without a key, however a keyed lock is required for the outside. Push button type locks are not allowed in the machine room door.
- ☐ Machine rooms shall have a minimum seven (7) feet clear headroom under all obstructions including and beams, light fixtures, etc.
- ☐ Machine room floor to be smooth and level
- ☐ Machine rooms shall have adequate ventilation to prevent overheating of the equipment. Ambient temperature in this room to be maintained between 60-90 degrees F at all times or as detailed on the submittal. Relative humidity is to not exceed 95% non-condensing.
- ☐ If situated on exterior wall, added precautions must be taken to prevent entry of blowing rain.
- ☐ Adequate lighting shall provide not less than ten (10) foot candles at floor level. The light switch shall be located on the lock jamb side of the access door.
- ☐ GFI receptacles shall be provided in the elevator machine room.
- ☐ All voids, holes, slots, etc in the machine room shall be grouted or pointed up to maintain the fire rating. All nails, snap-ties, form straps and wood shall be removed from the machine room walls, floor and ceiling.
- ☐ A disconnect switch for each elevator's main power supply shall be located within 18" of the strike side of the machine room door (see NEC 620-15). A 3'-6" clearance is required in front of the disconnect switch(es).
- ☐ The disconnect switch is to contain fusetrons (fuses that provide interrupting capacity and are marked "current limiting"). Fuse rejection clips must be furnished in the disconnect switch.
- ☐ A disconnecting means shall be provided for car lights, fan, car signals, viscosity control heaters and other equipment for each elevator. Disconnect must be lockable in the "off" position.
- ☐ Plastic electrical pipe cannot be used in the machine room or hoistway. Maximum length of flexible conduit cannot be longer than six (6) feet.
- ☐ Flexible conduit may not be used as a bond or ground to elevator equipment. A separate ground wire must be installed.

Figure 4.11 Elevator Machine Room Checklist.

SUBCONTRACTOR NEGOTIATION FORM

PROJECT

TRADE	SPECIFIED SECTIONS	DATE
Excavation and Site Work		
SUBCONTRACTOR	REPRESENTED BY	(AREA CODE) TELEPHONE NO.
BASE BID AMOUNT	ADDENDUM NO.	
ALTERNATES	UNIT PRICES	UNIT PRICES
(1)	(1)	(6)
(2)	(2)	(7)
(3)	(3)	(8)
(4)	(4)	(9)
(5)	(5)	(10)
SALES TAX	INSURANCE	

SCOPE OF WORK Including but not limited to the following:

This form must be completely filled out and signed by the subcontractor and the general contractor's representative to provide a record of this negotiation session

ITEM	YES	NO	EXPLANATION AND/OR COMMENTS
1. Clear and grub site, remove tree stumps from site. Provide all labor and materials to install and maintain silt fences until directed by GC to remove.			
2. Protect all trees to remain with sturdy barricades installed at the drip line.			
3. Topsoil: Strip and Stockpile on site Strip and truck off site Respread topsoil			
4. Dispose of unsuitable materials: Onsite as directed Offsite			
5. Remove all walks, curbs, paving, appurtenances, concrete foundation as directed by the contract documents. Dispose of offsite or dispose onsite at direction of GC.			
6. Site grading–cuts and fills			
7. Truck excess off site or Furnish Borrow as required			
8. Furnish compacted base materials as required for walks, paving, walls, curbs, equipment pads, etc			
9. Pit and trench excavation for all retaining walls, walks, curbs, pads.			

	FINAL AGREED AMOUNT
GENERAL CONTRACTOR'S REPRESENTATIVE SIGNATURE	SUBCONTRACTOR'S REPRESENTATIVE SIGNATURE

ORIGINAL

Figure 4.12 Subcontractor Form — Excavation and Site Work. *(continued on next page)*

CONTINUATION SHEET

SUBCONTRACTOR NEGOTIATION FORM

Page 2 of 2

ITEM	YES	NO	EXPLANATION AND/OR COMMENTS
10. Building excavation and backfill			
11. Interior building Pit and Trench excavation and backfill and compaction as required			
12. Dispose of excess: On-site as directed by GC: Off site			
13. Provide bedding and compaction for all interior underground utilities in accordance with the contract documents			
14. Provide and compact porous fill under slab on grade as specified in the contract documents			
15. Furnish and install all foundation drainage systems			
16. Rock excavation that is not rippable: Mass: Trench:			
17. Dewatering as required for all work included in the scope of this contract			
18. Provide all temporary barriers and barricades as required			
19. Provide and maintain all construction entrances with anti-tracking materials as required			
20. Subcontractor has received, read and accepts the subcontract agreement provided by the GC.			
21. Subcontractor will provide a detailed Schedule of Values for approval by the GC.			
22. Subcontractor will return a signed copy of the subcontract agreement within _____ days of receipt and cannot commence work until a fully executed agreement has been received by GC.			
See Excavation and Site Utilities Detail Sheets which may be attached to this interview form.			

Figure 4.12 *(continued from previous page)* Subcontractor Form — Excavation and Site Work.

Excavation and Site Work – Detail Sheet No. 1

- ❑ Sheeting- Steel Permanent
- ❑ Sheeting- Steel Temporary
- ❑ Accessories such as walers, diagonals, wall anchors, foot anchors
- ❑ Open wood sheathing
- ❑ Closed wood sheathing
- ❑ H Piles with wood cribbing
- ❑ Shoring as shown
- ❑ Bracing as shown or required
- ❑ Dewatering including all accessories, pumps, hoses, filtration
- ❑ Gravity run-off complete with piping, stone, and costs to restore

Well Point System

- ❑ Install and pump(s) and accessories
- ❑ Header and well point pipes- furnish, install, removal
- ❑ Jetting
- ❑ Drilling
- ❑ Pea gravel
- ❑ Regular shift
- ❑ Overtime shift

Foundation Underdrains

- ❑ Excavation and backfill
- ❑ Piping - Furnish and install
- ❑ Gravel fill

Figure 4.13 Excavation and Site Work Detail Sheet No. 1.

Exterior Mechanical and Electrical Utilities - Detail Sheet No. 2

Sanitary Sewer System

- ❑ Furnish and install all sanitary piping as required
- ❑ Excavation and backfill for sanitary piping and manholes
- ❑ Provide all bedding material as required
- ❑ Furnish and install sanitary manholes of type and in location required
- ❑ Furnish and install sanitary manhole frames and covers
- ❑ Furnish and install Lift Station, if required
- ❑ Furnish and install all valve pits as required
- ❑ Provide tie-in to existing sanitary system to include all permits and fees

Storm Sewer System

- ❑ Furnish and install storm sewer system piping as required
- ❑ Excavate/backfill for storm sewer piping, manholes, drain inlets, etc
- ❑ Provide all bedding material as required
- ❑ Provide tie into existing storm sewer system to include all permits and fees

Domestic Water System

- ❑ Furnish and install all piping as required
- ❑ Excavate and backfill using engineer approved bedding materials as required
- ❑ Furnish and install all valve pits as required
- ❑ Furnish and install all manholes as required
- ❑ Provide tie-in to existing potable water system to include all permits and fees
- ❑ Sanitize and disinfect system as required for acceptance by all authorities

Fire Protection System

- ❑ Furnish and install all piping as required
- ❑ Excavation and backfill for all piping and related structure
- ❑ Furnish and install all valve pits, manholes as required

Fuel Oil System for Building or Emergency Generator

- ❑ Excavate and backfill for piping supplied by others
- ❑ Excavate for underground tank supplied by others
- ❑ Furnish all labor and materials to install concrete pad with hold-down clamps
- ❑ Backfill with approved materials and compact as required

Electrical Service System

- ❑ Excavate, backfill for all conduit, encase in concrete as required. Compact as required.
- ❑ Provide and install all access pits, pull stations, vaults and manholes
- ❑ Furnish all precast appurtenances including transformer pads
- ❑ Underground telephone and data transmission cables
- ❑ Excavate and backfill for all tel/data communication conduits and direct burial
- ❑ Backfill and compact with materials approved by the engineer
- ❑ Provide and install all access pits, pull stations, vaults, manholes as required

Site Lighting

- ❑ Excavate and backfill for all cables
- ❑ Excavate and backfill for all site light supports and bases
- ❑ Provide labor and material for all concrete site light bases. Install anchor bolts, with templates prov electrical contractor
- ❑ Backfill for all cables with materials approved by the engineer
- ❑ All backfill to be compacted per specifications

Figure 4.14 Exterior Mechanical and Electrical Utilities Detail Sheet No. 2.

SUBCONTRACTOR NEGOTIATION FORM

PROJECT		
TRADE **Glass and Glazing**	SPECIFIED SECTIONS	DATE
SUBCONTRACTOR	REPRESENTED BY	(AREA CODE) TELEPHONE NO.
BASE BID AMOUNT	ADDENDUM NO.	
ALTERNATES (1)	UNIT PRICES (1)	UNIT PRICES (6)
(2)	(2)	(7)
(3)	(3)	(8)
(4)	(4)	(9)
(5)	(5)	(10)
SALES TAX	INSURANCE	

SCOPE OF WORK Including but not limited to the following:

This form must be completely filled out and signed by the subcontractor and the general contractor's representative to provide a record of this negotiation meeting.

ITEM	YES	NO	EXPLANATION AND/OR COMMENTS
Scope:			
1. Windows			
2. Curtain wall			
3. Entrances			
4. Storefront			
5. Exterior doors			
6. Interior doors			
7. Spandrel panels			
8. Interior partitions			
9. Interior hollow metal frames/ Hollow metal doors/ borrowed light frames (delete item(s) not applicable)			

	FINAL AGREED AMOUNT
GENERAL CONTRACTOR'S REPRESENTATIVE SIGNATURE	SUBCONTRACTOR'S REPRESENTATIVE SIGNATURE

ORIGINAL

Figure 4.15 Subcontractor Form — Glass and Glazing. *(continued on next page)*

CONTINUATION SHEET

SUBCONTRACTOR NEGOTIATION FORM Page 2 of 2

ITEM	YES	NO	EXPLANATION AND/OR COMMENTS
General: 10. Shop drawings, samples. Mock-ups required, warranties/ guarantees (provide bond for guarantees in excess of one year).			
11. Contractor has reviewed complete drawings and specifications including general and special conditions. Contractor has inspected site (if applicable).			
12. Contractor includes all escalation costs for the life of the contract. Contractor includes all hoisting, lifting, scaffolding for their work.			
13. Contractor will provide all engineering and layout for their work.			
14. Insurance coverage is per specification. Bond will be provided, if required.			
15.Contractor agrees to abide by all OSHA regulations and general contractor's safety plan.			
16. All local, state, and federal taxes are included. If tax exempt, are all applicable taxes not included?			
17. Glass Types(check all types included): Clear polished – thickness			
18. Mirrors- specify location, number, size, thickness.			
19. Glass shelves- specify location, number, size.			
20. State labor rates, including burden for all tradesmen being employed for this project.			
21. Specify percent overhead and profit added to all labor and material costs for extra work.			
22. Contractor agrees to protect all work until accepted and replace any damaged work at no cost to the general contractor or the owner.			
23. Contractor agrees to remove all debris from the site on a daily basis and keep their work area clean. This space to be used for any special conditions			

Figure 4.15 *(continued from previous page)* Subcontractor Form — Glass and Glazing.

SUBCONTRACTOR NEGOTIATION FORM

PROJECT

TRADE	SPECIFIED SECTIONS	DATE
HVAC		
SUBCONTRACTOR	REPRESENTED BY	(AREA CODE) TELEPHONE NO.
BASE BID AMOUNT	ADDENDUM NO.	
ALTERNATES	UNIT PRICES	UNIT PRICES
(1)	(1)	(6)
(2)	(2)	(7)
(3)	(3)	(8)
(4)	(4)	(9)
(5)	(5)	(10)
SALES TAX	INSURANCE	

SCOPE OF WORK Including but not limited to the following:

This form must be completely filled out and signed by subcontractor and the general contractor's representative to record this negotiation meeting.

ITEM	YES	NO	EXPLANATION AND/OR COMMENTS
1. Furnish all labor, materials, and equipment to complete the HVAC installation in strict accordance with the contract documents.			
2. The subcontractor has received, read, and accepted the contract specifications including the general and special conditions.			
3. Subcontractor to furnish starters for all their equipment, U.O.N and coordinate electrical characteristics of all equipment with the electrical subcontractor. Confirm results to GC.			
4. Subcontractor to submit duct shop drawings for review and comment at a subcontractor's coordination meeting.			
5. Cutting within the jurisdiction of the subcontractor; patching of surfaces due to errors of subcontractor.			
6. Control wiring and all low voltage wiring.			
7. Louvers, interior and exterior as required for completion of subcontractor's work.			
8. Diffusers and grills, VAV boxes, transfer grills and other such terminal devices.			
9. Condensate lines piped to nearest drain.			

	FINAL AGREED AMOUNT
GENERAL CONTRACTOR'S REPRESENTATIVE SIGNATURE	SUBCONTRACTOR'S REPRESENTATIVE SIGNATURE

ORIGINAL

Figure 4.16 Subcontractor Form — HVAC. *(continued on next page)*

CONTINUATION SHEET

SUBCONTRACTOR NEGOTIATION FORM

Page 2 of 2

ITEM	YES	NO	EXPLANATION AND/OR COMMENTS
10. Curbs, dunnage, supports for roof top equipment- furnish and install			
11. Steam distribution system as required			
12. Refrigerant lines as required			
13. Condensor water circulation system Make-up water system			
14. Pipe and duct insulation (interior /exterior) as specified.			
15. Sleeves and inserts - furnish and install			
16. Pumps, circulating water, condensate, vacuum - furnish and install			
17. Induction units			
18. Cooling Tower – type –furnish, rig, install, run-up test			
19. Pipe, duct, equipment identification markers, color coding, plastic equipment plaques			
20. Test, adjust, balance. Submit report for engineer's review and approval			
21. Expansion tanks - furnish and install			
22. Emergency generator exhaust piping			
24. Black iron duct for kitchen exhaust - furnish and install 25. Vent piping- furnish and install			

Figure 4.16 *(continued from previous page)* Subcontractor Form — HVAC.

SUBCONTRACTOR NEGOTIATION FORM

Page 1 of 2

PROJECT		
TRADE **Masonry**	SPECIFIED SECTIONS	DATE
SUBCONTRACTOR	REPRESENTED BY	(AREA CODE) TELEPHONE NO.
BASE BID AMOUNT	ADDENDUM NO.	
ALTERNATES (1)	UNIT PRICES (1)	UNIT PRICES (6)
(2)	(2)	(7)
(3)	(3)	(8)
(4)	(4)	(9)
(5)	(5)	(10)
SALES TAX	INSURANCE	

SCOPE OF WORK Including but not limited to the following:

This form is to be completely filled out and signed by the subcontractor and the general contractor's representative to record the negotiation meeting.

ITEM	YES	NO	EXPLANATION AND/OR COMMENTS
1. Furnish samples, sample panel, and/or mock-up per specifications.			
2. Build in work of other trades as required.			
3. Scaffolding and hoisting for own work. Scaffold planks to be cleaned of excess at end of each work day.			
4. Layout and engineering for own work			
5. Bracing and protecting of all work			
6. Bond pattern, joint width and type (tooled, raked, flush, etc) per specifications			
7. Mortar type, color (integral, not added) and mixes as specified			
8. Furnish and install all masonry reinforcement and anchoring devices in strict accordance with the specifications.			
9. Face brick-size, allowance, manufacturer			

	FINAL AGREED AMOUNT
GENERAL CONTRACTOR'S REPRESENTATIVE SIGNATURE	SUBCONTRACTOR'S REPRESENTATIVE SIGNATURE

ORIGINAL

Figure 4.17 Subcontractor Form — Masonry. *(continued on next page)*

CONTINUATION SHEET

SUBCONTRACTOR NEGOTIATION FORM

Page 2 of 2

ITEM	YES	NO	EXPLANATION AND/OR COMMENTS
10. Special shapes (list)			
11.CMU –types (list)			
12. Sills, copings, decorative pieces- furnish and install			
13. Control joints, expansion joints; furnish- install.			
14. Grout fill of CMU and bond beams to include furnishing and installation of all rebars			
15. Weeps, pea gravel cavity wall drainage to be keep free of all debris, excess mortar			
16. Furnish and install all through wall flashings and flashing materials supplied by others.			
17. Lintels - furnish, receive, unload and install all precast lintels. Install steel lintels supplied by others.			
18. Install rigid insulation by others in cavity walls as required.			
19. Reglets as required			
20. Non–acid clean-down, additional wash downs required to eliminate effluorescence within 45 days after completion			
21. Debris to be placed in dumpster as directed by general contractor			
22. Site Masonry (list)			
23. Subcontractor has received, read, and accepts general contractor's subcontract agreement.			
24. Subcontractor has received, read and accepts general contractor's safety program.			

Figure 4.17 *(continued from previous page)* Subcontractor Form — Masonry.

SUBCONTRACTOR NEGOTIATION FORM

PROJECT

TRADE	SPECIFIED SECTIONS	DATE
Metal Deck		
SUBCONTRACTOR	REPRESENTED BY	(AREA CODE) TELEPHONE NO.
BASE BID AMOUNT	ADDENDUM NO.	
ALTERNATES	UNIT PRICES	UNIT PRICES
(1)	(1)	(6)
(2)	(2)	(7)
(3)	(3)	(8)
(4)	(4)	(9)
(5)	(5)	(10)
SALES TAX	INSURANCE	

SCOPE OF WORK Including but not limited to the following:

This form is to be completely filled out and signed by the subcontractor and the general contractor's representative to provide a record of this negotiation meeting.

ITEM	YES	NO	EXPLANATION AND/OR COMMENTS
1. Metal floor deck: furnish, hoist, erect, install Furnish & install shear studs Quantity check_____ SF Breakout price:$_____			
2. Metal roof deck: Painted/ Galvanized Furnish, hoist, erect, install Quantity check_____ SF Breakout price: $_________			
3. Materials: Sizes, gauges, finishes, configurations as specified			
4. Closures, edge angles, and pour stops as required			
5. Drain accessories			
6. Reinforcement materials at column abutments			
7. Reinforcements at all required openings per A, S, and MEP drawings			
8. Caps, covers, and flashings as required			
9. Sump recesses – Furnish and install as required			

	FINAL AGREED AMOUNT
GENERAL CONTRACTOR'S REPRESENTATIVE SIGNATURE	SUBCONTRACTOR'S REPRESENTATIVE SIGNATURE

ORIGINAL

Figure 4.18 Subcontractor Form — Metal Deck. *(continued on next page)*

CONTINUATION SHEET

SUBCONTRACTOR NEGOTIATION FORM

Page 2 of 2

ITEM	YES	NO	EXPLANATION AND/OR COMMENTS
10. Hanger tabs, slots and/or clips – furnish and install			
11. Other inserts – furnish and install			
12. Includes setting metal deck resting on concrete and/or masonry as required			
13. Shoring and bracing as required			
14. Cut openings in places and manner as specified or shown on all drawings, including mechanical openings. Reinforce as required			
15. Provide temporary protection around work, specifically at all openings - cover or barricade			
16. Composite deck as required			
17. Touch-up all surfaces as required			
18. Subcontractor to provide temporary power, at their cost, if not available by GC			
19. Subcontractor has reviewed complete plans and specs including general and special requirements and accepts those pertaining to their work.			
20. Subcontractor has received, read, and accepts general contractor's safety program and all applicable OSHA requirements.			
21. Subcontractor is to return subcontract agreement within ____ days of receipt, if an award is made. No work can commence on site unless a signed agreement is received by GC.			
22. Subcontractor to provide site logistic plan to GC for approval, showing lay-down area and site access required for their work.			

Figure 4.18 *(continued from previous page)* Subcontractor Form — Metal Deck.

SUBCONTRACTOR NEGOTIATION FORM

PROJECT

TRADE Metal Framing & Drywall	SPECIFIED SECTIONS	DATE
SUBCONTRACTOR	REPRESENTED BY	(AREA CODE) TELEPHONE NO.
BASE BID AMOUNT	ADDENDUM NO.	
ALTERNATES	UNIT PRICES	UNIT PRICES
(1)	(1)	(6)
(2)	(2)	(7)
(3)	(3)	(8)
(4)	(4)	(9)
(5)	(5)	(10)
SALES TAX	INSURANCE	

SCOPE OF WORK Including but not limited to the following:

This form to be filled out and signed by subcontractor and the general contractor's representative to record this negotiation meeting.

ITEM	YES	NO	EXPLANATION AND/OR COMMENTS
1. Include all necessary labor, materials, equipment, tools, hoisting and lifting equipment to install metal framing and drywall work per the contract documents (list)			
2. Perform all required cutting, patching, framing, drilling, tapping, bending to install your work			
3. Provide materials as specified or required to conform to all STC, fire ratings, and applicable building codes			
4. Furnish and install floor and ceiling runner fabricated of __________ gauge_______metal			
5. Furnish and install wall and ceiling furring channels of ________ gauge and________metal			
6. Furnish and install steel stud fabricated of____ &____ gauge and _____ metal			
7. Furnish and install all specified or required corner reinforcement and trim pieces			
8. Furnish and install all required bolts, screws, and fastening devices required			
9. Furnish and install all isolation joints between partitions and structure as required			
10. Perform welding of structural studs as required			

FINAL AGREED AMOUNT

GENERAL CONTRACTOR"S REPRESENTATIVE SIGNATURE	SUBCONTRACTOR'S REPRESENTATIVE. SIGNATURE

ORIGINAL

Figure 4.19 Subcontractor Form — Metal Framing and Drywall. *(continued on next page)*

CONTINUATION SHEET

SUBCONTRACTOR NEGOTIATION FORM

Page 2 of 3

ITEM	YES	NO	EXPLANATION AND/OR COMMENTS
11. Furnish and install additional stud framing as required or specified at corners, openings, door, frames, handrails, grab bars, toilet accessories			
12. Furnish and install gypsum board of thickness and type required - fire-rated, moisture- resistant, exterior sheathing, prefinished as required			
13. Furnish and install shaftwall liner board as required			
14. Furnish and install metal-edged gypsum plank as required			
15. Furnish and install specified insulation between studs			
16. Furnish and install rigid insulation at ______________________			
17. Furnish and install all sealant and/or caulking as specified or required or as directed at________________ locations			
18. Perform all required taping of joints with individual number of coats as specified. Sand acceptable to painter			
19. Unload, distribute and install the following: a. Hollow metal frames, doors, borrowed lites b. Lead lined sheet rock c. Access doors d. Built-in items e. Other ____________________________			
20. Furnish and install additional framing required for support and back-up of wall and ceiling hung items ,handrails, toilet accessories, light fixtures			
21. Provide cut-out for electrical boxes, switches, etc. tight to the rough-in			
22. Furnish and install special corner guards			
23. Furnish and install soffits, drapery pockets, etc.			

Figure 4.19 *(continued from previous page)* Subcontractor Form — Metal Framing and Drywall.

CONTINUATION SHEET

SUBCONTRACTOR NEGOTIATION FORM

ITEM	YES	NO	EXPLANATION AND/OR COMMENTS
24. Nails, screws, and glue, as specified			
25. Furnish and install temporary protection for all alteration/renovation work.			
26. Rubbish removal, on a daily basis, to a dumpster designated by general contractor. All rubbish to be broken into small pieces.			
27. Hoisting and scaffolding for own work			
28. Subcontractor has read and accepted general contractor's safety plan.			
29. Subcontractor has read and accepted terms and conditions of general contractor's subcontract agreement form.			
30. Insurance as required________________ Payment and Performance bond__________ EEO Requirements accepted_____________ Taxes Included _____ or Tax Exempt______			
31. This space reserved for unit prices			
32. Subcontractor's OH&P on change orders including second tier subcontractors.			
33. Subcontractor has reviewed project construction schedule and accepts start and duration of work for their trade(s).			
34. If awarded subcontract agreement, subcontractor must return within 5 days after receipt.			
35. Subcontractor will not be permitted on site without executed subcontract agreement and insurance certificates as specified.			
36. Subcontractor to submit detailed Schedule of Values for their work to be approved by the Project Manager.			

Figure 4.19 *(continued from previous page)* Subcontractor Form — Metal Framing and Drywall.

SUBCONTRACTOR NEGOTIATION FORM

Page 1 of 2

PROJECT		
TRADE **Miscellaneous Metals**	SPECIFIED SECTIONS	DATE
SUBCONTRACTOR	REPRESENTED BY	(AREA CODE) TELEPHONE NO.
BASE BID AMOUNT	ADDENDUM NO.	
ALTERNATES (1)	UNIT PRICES (1)	UNIT PRICES (6)
(2)	(2)	(7)
(3)	(3)	(8)
(4)	(4)	(9)
(5)	(5)	(10)
SALES TAX	INSURANCE	

SCOPE OF WORK Including but not limited to the following:

This form must be completely filled out and signed by the subcontractor and general contractor's representative to provide a record of this negotiation meeting.

ITEM	YES	NO	EXPLANATION AND/OR COMMENTS
1. All miscellaneous metal items are to be furnished and installed, unless otherwise noted.			
2. Galvanized as required/ Shop primed as required			
3. Bituminous coating as required			
4. All accessories such as bolts, screws, clips, anchors, etc to fabricate and install work			
5. Furnish inserts/anchoring devices set into concrete/ masonry for attachment to miscellaneous metal items			
6. Furnish inserts and/or anchoring devices for other trades as specified			
7. Shop assemble to greatest extent possible			
8. Field measurements as required			
9. Coordinate delivery, setting drawings, diagrams, templates, instructions, and directions for inserts, anchors, bolts or anchorages set by others			

	FINAL AGREED AMOUNT
GENERAL CONTRACTOR'S SIGNATURE	SUBCONTRACTOR'S REPRESENTATIVE SIGNATURE

ORIGINAL

Figure 4.20 Subcontractor Form — Miscellaneous Metals. *(continued on next page)*

CONTINUATION SHEET

SUBCONTRACTOR NEGOTIATION FORM

Page 2 of 2

ITEM	YES	NO	EXPLANATION AND/OR COMMENTS
10. Subcontractor has accepted GC's safety program and all OSHA safety regulations			
11. Subcontractor to provide all temporary power unless general contractor agrees to provide			
12. If lay-down area is required, provide drawing for general contractor's approval			
13. Specified manufacturers or "Or Equal" only (burden of proof on Subcontractor)			
14. Insurance certificates to be furnished to meet limits as set forth in the contract specifications			
15. Subcontractor has received, read, and accepts terms/conditions of subcontractor agreement			
16. A payment and performance bond will be forthcoming, if requested			
17. For additional work ________% overhead and _____% profit will apply			
18. Subcontractor has received all dwgs, including A,M,E,P and includes all related work			
19. Subcontractor has received GC's base line schedule and accepts portion for their trade			
20. All items requiring factory finish to be in strict accordance with specification requirements			
21. Submit shop drawing schedule and delivery of each item upon receipt of approval			
22. Acknowledge attached detail list(s) to be appended to this form			

Figure 4.20 *(continued from previous page)* Subcontractor Form — Miscellaneous Metals.

Miscellaneous Metal - Detail Lists

Steel Stairs

- ❑ Steel-framed stairs including metal framing, hangers, columns, struts, clips, brackets, bearing plates, treads, risers, platforms
- ❑ Temporary supports
- ❑ Metal pan units as specified or shown
- ❑ Metal safety nosing as specified or shown
- ❑ Steel floor plate treads as specified or shown
- ❑ Diamond plate as specified or shown
- ❑ Grating construction
- ❑ Railings of size - amount, gauge, and material
- ❑ Kick plates
- ❑ Fabrication as specified or shown
- ❑ Open riser stairs

Handrails and Railings

- ❑ Material of type, grades, finishes, weights, construction, tolerances as specified
- ❑ Steel pipe
- ❑ Galvanized pipe
- ❑ Gray iron castings
- ❑ Malleable iron castings
- ❑ Stainless steel pipe
- ❑ Aluminum pipe
- ❑ Other types of railings

  a.________________

  b.________________

  c________________

Expansion Joints

- ❑ Materials of type, grade, finish, weight, construction, and tolerances as required
- ❑ Floor expansion joints - angles with anchors
- ❑ Floor expansion joint covers
- ❑ Wall expansion joint covers
- ❑ Ceiling expansion joint covers
- ❑ All accessories of equivalent construction or grade as specified for a complete installation

Wall, Floor, and/or Ceiling Supports and Reinforcements for:

- ❑ Handrails
- ❑ Grab bars
- ❑ Toilet accessories
- ❑ Toilet partitions
- ❑ Television sets- wall or ceiling mounted
- ❑ Miscellaneous hospital equipment
- ❑ Wall-mounted equipment

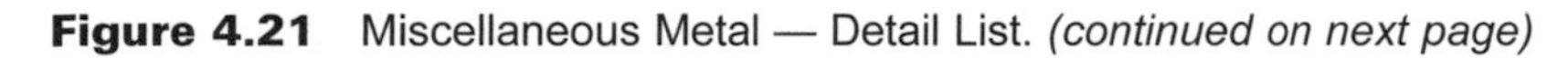

Figure 4.21 Miscellaneous Metal — Detail List. *(continued on next page)*

- ❑ Exhaust hoods
- ❑ Roof openings
- ❑ Knee wall partitions
- ❑ Countertops- toilet
- ❑ Countertops- kitchens
- ❑ Wall shelving

Miscellaneous Supports for Precast Concrete, Cast Stone, Limestone, Granite, etc.

- ❑ Bolts
- ❑ Clip angles
- ❑ Struts
- ❑ Bracing
- ❑ Relieving angles
- ❑ Inserts
- ❑ Wedges
- ❑ Tie backs
- ❑ Support framing

Miscellaneous Supports for Curtain Wall, Storefront, Window Wall, etc.

- ❑ Bolts
- ❑ Clip angles
- ❑ Struts
- ❑ Bracing
- ❑ Head supports
- ❑ Jamb supports
- ❑ Sill supports
- ❑ Inserts, wedges

Exterior Miscellaneous Metal Requirements

- ❑ Manhole, catch basin, trench drain, drain inlet frames
- ❑ Exterior ladder rungs for manholes, catch basins, drain inlets
- ❑ Exterior access hatches, frames, and rungs for mechanical, electrical, and other trades
- ❑ Bench supports
- ❑ Wall rail supports
- ❑ Abrasive nosing for concrete stairs
- ❑ Abrasive nosings for door sills
- ❑ Metal saddles
- ❑ Metal thresholds
- ❑ Roof hatches
- ❑ Roof scuttles
- ❑ Catwalks – framing, stairs, railings, gratings
- ❑ Dock bumpers

Figure 4.21 *(continued from previous page)* Miscellaneous Metal — Detail List.

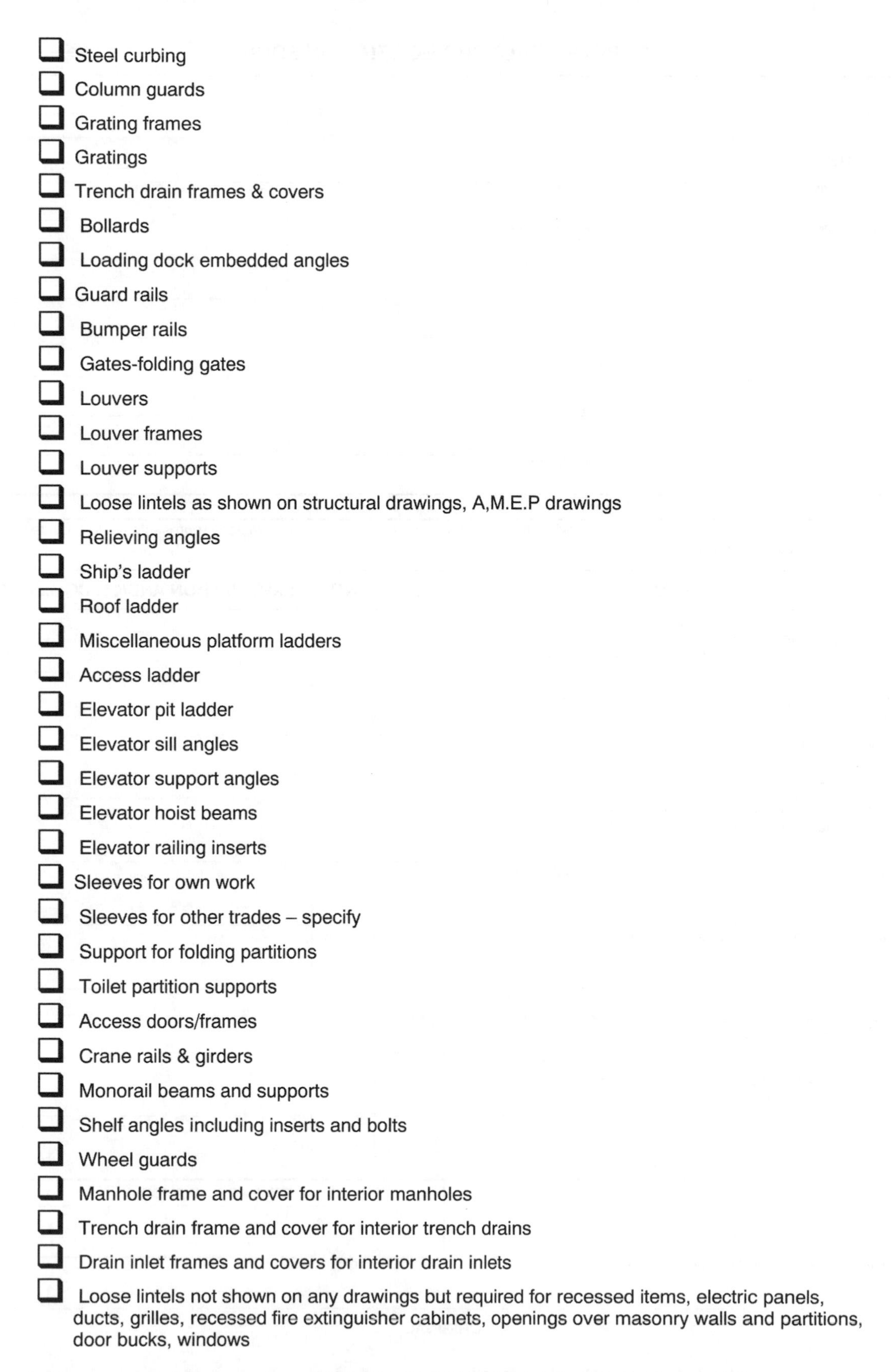

- ❑ Steel curbing
- ❑ Column guards
- ❑ Grating frames
- ❑ Gratings
- ❑ Trench drain frames & covers
- ❑ Bollards
- ❑ Loading dock embedded angles
- ❑ Guard rails
- ❑ Bumper rails
- ❑ Gates-folding gates
- ❑ Louvers
- ❑ Louver frames
- ❑ Louver supports
- ❑ Loose lintels as shown on structural drawings, A,M.E.P drawings
- ❑ Relieving angles
- ❑ Ship's ladder
- ❑ Roof ladder
- ❑ Miscellaneous platform ladders
- ❑ Access ladder
- ❑ Elevator pit ladder
- ❑ Elevator sill angles
- ❑ Elevator support angles
- ❑ Elevator hoist beams
- ❑ Elevator railing inserts
- ❑ Sleeves for own work
- ❑ Sleeves for other trades – specify
- ❑ Support for folding partitions
- ❑ Toilet partition supports
- ❑ Access doors/frames
- ❑ Crane rails & girders
- ❑ Monorail beams and supports
- ❑ Shelf angles including inserts and bolts
- ❑ Wheel guards
- ❑ Manhole frame and cover for interior manholes
- ❑ Trench drain frame and cover for interior trench drains
- ❑ Drain inlet frames and covers for interior drain inlets
- ❑ Loose lintels not shown on any drawings but required for recessed items, electric panels, ducts, grilles, recessed fire extinguisher cabinets, openings over masonry walls and partitions, door bucks, windows

Figure 4.21 *(continued from previous page)* Miscellaneous Metal — Detail List.

SUBCONTRACTOR NEGOTIATION FORM

Page 1 of 2

PROJECT		
TRADE **Painting**	SPECIFIED SECTIONS	DATE
SUBCONTRACTOR	REPRESENTED BY	(AREA CODE) TELEPHONE NO.
BASE BID AMOUNT	ADDENDUM NO.	
ALTERNATES (1)	UNIT PRICES (1)	UNIT PRICES (6)
(2)	(2)	(7)
(3)	(3)	(8)
(4)	(4)	(9)
(5)	(5)	(10)
SALES TAX	INSURANCE	

SCOPE OF WORK Including but not limited to the following:

This form must be completely filled out and signed by the subcontractor and the general contractor's representative to provide a record of this negotiation meeting.

ITEM	YES	NO	EXPLANATION AND/OR COMMENTS
1. Include all labor, materials, equipment, hoisting, and scaffolding to complete painting work in accordance with the all contract documents- A,S.MEP drawings and specifications.			
2. Structural steel & metal deck painting/touch up and miscellaneous metal (interior and exterior) painting/touch-up (List)			
3. Hollow metal work, frames, doors, borrowed lites			
4. Millwork finishing, cabinetry, built-ins, wood railings, running trim, miscellaneous architectural woodwork			
5. Drywall- subcontractor to inspect and accept substrate prior to commencing work			
6. Masonry, where required, to include block filler, special coatings			
7. Mechanical work – piping, pipe covering, ductwork, hangers, supports, equipment, color coding, as required by all applicable spec section(s)			
8. Electrical work – conduit, hangers, junction & pullboxes, panelboards, equipment, color coding, as required by all applicable spec sections			
9. Wood doors, frames and trim			

	FINAL AGREED AMOUNT
GENERAL CONTRACTOR'S REPRESENTATIVE SIGNATURE	SUBCONTRACTOR'S REPRESENTATIVE SIGNATURE

ORIGINAL

Figure 4.22 Subcontractor Form — Painting. *(continued on next page)*

CONTINUATION SHEET

SUBCONTRACTOR NEGOTIATION FORM Page 2 of 2

ITEM	YES	NO	EXPLANATION AND/OR COMMENTS
10. Materials and colors as specified. What manufacturer will be used? Provide samples and mock-ups as required.			
11. Number of coats specified for each substrate. List.			
12. Preparation of surfaces to include putty and spackle as required to make them acceptable to the architect			
13. Finishes on wood and metal surfaces to be sanded between coats to assure smoothness and adhesion of prior coat			
14. Mill thickness on all coats to be as specified			
15. All work in strict accordance with manufacturer's instructions			
16. Special coatings - glazed wall coating, fire resistant, epoxy			
17. Wall coverings - manufacturer, thickness, weight, roll width, per specifications			
18. Application: Roll, brush, spray. Identify for each substrate type.			
19. Protection as required for floors, walls, ceilings, diffusers, convectors, installed equipment. All masking of surfaces as required.			
20. Clean up, remove masking, deposit daily in dumpster in location provided by GC			
21. Subcontractor has received complete set of drawings and specifications including general and special conditions and assumes responsibility for all requirements pertaining to their			
22. Subcontractor to repair all defective and rejected work as directed by the Architect			
23. Additional work percentage: ____% overhead ____% profit 24. Subcontractor has received, read, and accepts general contractor's safety program.			

Figure 4.22 *(continued from previous page)* Subcontractor Form — Painting.

SUBCONTRACTOR NEGOTIATION FORM

Page 1 of 2

PROJECT		
TRADE Plumbing	SPECIFIED SECTIONS	DATE
SUBCONTRACTOR	REPRESENTED BY	(AREA CODE) TELEPHONE NO.
BASE BID AMOUNT	ADDENDUM NO.	
ALTERNATES (1)	UNIT PRICES (1)	UNIT PRICES (6)
(2)	(2)	(7)
(3)	(3)	(8)
(4)	(4)	(9)
(5)	(5)	(10)
SALES TAX	INSURANCE	

SCOPE OF WORK Including but not limited to the following:

This form must be completely filled out and signed by the subcontractor and the general contractor's representative to provide a record of this negotiation meeting.

ITEM	YES	NO	EXPLANATION AND/OR COMMENTS
1. Include all necessary labor, materials, and equipment as required to install all plumbing work in accordance with the contract documents.			
2. Subcontractor has reviewed specification section _______ and general and special conditions and agrees with all requirements pertaining to their trade.			
3. Insurance certificates per contract specification will be provided. A payment and performance bond, if required, will be furnished.			
4. Subcontractor has reviewed (company's) subcontract agreement and agrees to its terms.			
5. Subcontractor will provide a detailed Schedule of Values for its work in form and content approved by the general contractor.			
6. Subcontractor has reviewed the construction schedule and accepts the portion pertaining to their trade.			
7. Subcontractor to provide all layout work including line and grade as required.			
8. Subcontractor will perform all cutting and patching in a manner acceptable by the General Contractor.			
9. Subcontractor will obtain and pay for all permits required for their work including all connection charges.			

	FINAL AGREED AMOUNT
GENERAL CONTRACTOR'S REPRESENTATIVE SIGNATURE	SUBCONTRACTOR'S REPRESENTATIVE SIGNATURE

ORIGINAL

Figure 4.23 Subcontractor Form — Plumbing. *(continued on next page)*

CONTINUATION SHEET

SUBCONTRACTOR NEGOTIATION FORM Page 2 of 2

ITEM	YES	NO	EXPLANATION AND/OR COMMENTS
10. Subcontractor will provide all trenching, backfilling, and concrete encasement required for their work in a manner acceptable to the general contractor.			
11. Subcontractor to furnish and install all wall, floor, and ceiling access doors required for their work.			
12. Subcontractor to provide a shop drawing submittal list to include submission dates and delivery of materials/equipment upon receipt of architect's approval.			
13. Subcontractor responsible for coordination of all of their electrically powered equipment with the project's electrical subcontractor.			
14. Subcontractor to provide temporary water for trades as directed by the general contractor.			
15. Subcontractors to remove their rubbish and debris, on a daily basis, to a dumpster provided by the general contractor.			
16. Subcontractor to locate and provide all required anchor bolts for embedment by others.			
17. Subcontractor to have a qualified, experienced full-time project manager or superintendent on site – as required by the general contractor.			
18. Subcontractor to provide final connection to all "contract" equipment supplied by others.			
19. Subcontractor to install all toilet accessories supplied by others.			
20. Additional work percentages:_____% profit, ___% overhead			
21. Subcontractor has received, read, and accepts the company's safety program.			
22. The subcontractor is to return their subcontract agreement, if awarded, within _____ days. No work can commence until a signed contract has been received and accepted by the general contractor.			

Figure 4.23 *(continued from previous page)* Subcontractor Form — Plumbing.

SUBCONTRACTOR NEGOTIATION FORM

Page 1 of 2

PROJECT		
TRADE **Precast Concrete**	SPECIFIED SECTIONS	DATE
SUBCONTRACTOR	REPRESENTED BY	(AREA CODE) TELEPHONE NO.
BASE BID AMOUNT	ADDENDUM NO.	
ALTERNATES (1)	UNIT PRICES (1)	UNIT PRICES (6)
(2)	(2)	(7)
(3)	(3)	(8)
(4)	(4)	(9)
(5)	(5)	(10)
SALES TAX	INSURANCE	

SCOPE OF WORK Including but not limited to the following:

This form must be completely filled out and signed by the subcontractor and the general contractor's representative to provide a record of this negotiation meeting.

ITEM	YES	NO	EXPLANATION AND/OR COMMENTS
1. Include all labor, material, equipment, tools, and appurtenances to furnish and install precast concrete per the contract documents, including hoisting and scaffolding			
2. The scope of the work includes furnishing and erecting: architectural precast, structural precast, miscellaneous items such as site equipment pads, etc.			
3. Subcontractor to submit shop drawings prepared by__________, within ____ days of award. Delivery______ days after approval			
4. The subcontractor has reviewed complete drawings, including A,S,MEP and Civil, and all specs including general and special conditions.			
5. The subcontractor has received, read, and accepts the terms and conditions of the subcontract agreement.			
6. Subcontractor to provide samples and mock-up (if applicable), including finish and color, for approval			
7. Subcontractor to provide site logistics plan for approval to include lay-down area and required site access			
8. The following section(s) of the exterior wall will be left out until erection has finished: ______________________________			
9. Furnish and install all caulking or precast and precast to adjacent surfaces. Caulk all joints in precast plank, if furnished.			

	FINAL AGREED AMOUNT
GENERAL CONTRACTOR'S REPRESENTATIVE SIGNATURE	SUBCONTRACTOR'S REPRESENTATIVE SIGNATURE

ORIGINAL

Figure 4.24 Subcontractor Form — Precast Concrete. *(continued on next page)*

CONTINUATION SHEET

SUBCONTRACTOR NEGOTIATION FORM Page 2 of 2

ITEM	YES	NO	EXPLANATION AND/OR COMMENTS
10. Furnish and install all waterstops required at precast and adjacent surfaces			
11. Furnish and install all rebars for the connection of precast panels to masonry walls			
12. Build in or prepare precast to receive anchors or finishes supplied by others			
13. Cast panels with all box-outs and sleeved openings as required			
14. Furnish and install all required shims, reglets, clip angles, welding, bolts, dowels, anchors, and inserts required for the erection of precast			
15. The cost for additional crane rental, with operator, over and above that required to erect the precast is $_____/ (hour), (day), (week).			
16. Layout and engineering for own work - subcontractor has visited and inspected the site.			
17. Subcontractor has received, read, and accepts general contractor's safety program to include all OSHA requirements.			
18. Additional work percentages:____ % profit, ____% overhead.			
19. Subcontractor will repair/replace damaged precast members as directed by, and to the satisfaction of, the architect.			
20. Precast Checklist: Architectural: (1) wall panels (2) coping (3) soffits (4) sills (5) column covers (6) sun screens (7) others__________________			
21. Precast Checklist: Structural: (1) floor deck (2) columns (3) beams (4) girders (5) walls (6) stairs (7) others __________________________			
22. To facilitate jobsite mobilization and access to areas of the building by other trades, subcontractor agrees to sequence work as directed by the general contractor.			
23. Misc. items			

Figure 4.24 *(continued from previous page)* Subcontractor Form — Precast Concrete.

SUBCONTRACTOR NEGOTIATION FORM

PROJECT		
TRADE **Resilient Flooring**	SPECIFIED SECTIONS	DATE
SUBCONTRACTOR	REPRESENTED BY	(AREA CODE) TELEPHONE NO.
BASE BID AMOUNT	ADDENDUM NO.	
ALTERNATES (1)	UNIT PRICES (1)	UNIT PRICES (6)
(2)	(2)	(7)
(3)	(3)	(8)
(4)	(4)	(9)
(5)	(5)	(10)
SALES TAX	INSURANCE	

SCOPE OF WORK Including but not limited to the following:

This form must be completely filled out and signed by the subcontractor and the general contractor's representative to provide a record of this negotiation session.

ITEM	YES	NO	EXPLANATION AND/OR COMMENTS
1. Vinyl composition tile in size, thickness, color, texture in accordance with contract requirements. All materials from one manufacturer, U.O.N.			
2. Resilient sheet goods in thickness, color, pattern, texture per contract documents			
3. Patterns of VCT in accordance with architectural drawings and architect approval			
4. Adhesives per specification and installed in strict accordance with manufacturer's instructions			
5. Subcontractor to inspect and accept substrate prior to commencement of work. Flash patch as required. Moisture test for concrete to be subcontractor's obligation			
6. Vinyl base in color, size, and manufacturer to be in accordance with contract requirements			
7. Division strips, transition, and edging strips, as required			
8. Surfaces to be broom cleaned or vacuumed prior to start of work			
9. Cut tile to fit neatly around all fixtures and equipment. Cuts less than 4 inches will not be accepted.			

	FINAL AGREED AMOUNT
GENERAL CONTRACTOR'S REPRESENTATIVE SIGNATURE	SUBCONTRACTOR'S REPRESENTATIVE SIGNATURE

ORIGINAL

Figure 4.25 Subcontractor Form — Resilient Flooring. *(continued on next page)*

CONTINUATION SHEET

SUBCONTRACTOR NEGOTIATION FORM

ITEM	YES	NO	EXPLANATION AND/OR COMMENTS
10. Subcontractor to confirm that room environment meets manufacturer's requirements prior to commencement of work.			
11. Install tile in toe spaces, door reveals, closets, and storage areas adjacent to rooms being finished.			
12. Remove excessive adhesive from surrounding surfaces and clean thoroughly.			
13. Resilient flooring to be installed under all cabinetry and equipment in rooms where this flooring is required. Coordinate with appropriate trades.			
14. Hoisting and distribution of materials is included.			
15. All defective tiles/sheet goods are to be repaired and subject to architect approval.			
16. At recessed base detail, Subcontractor shall trim base to fit due to irregularities in the floor slab.			
17. Preformed vinyl base corners or back-scored base to be provided to fit corners tightly to wall or equipment base.			
18. Clean floors, etc. prior to final inspection by architect.			
19. Wax and buff per manufacturer's instructions.			
20. Protect finished floors with heavy kraft paper or similar material. Remove all debris to dumpster in location as directed by GC.			
21. Subcontractor has received, read, and accepts terms of subcontract agreement.			
22. Subcontractor has received, read, and accepts terms of general contractor's safety program.			

Figure 4.25 *(continued from previous page)* Subcontractor Form — Resilient Flooring.

SUBCONTRACTOR NEGOTIATION FORM

Page 1 of 2

PROJECT		
TRADE **Roofing & Sheet Metal**	SPECIFIED SECTIONS	DATE
SUBCONTRACTOR	REPRESENTED BY	(AREA CODE) TELEPHONE NO.
BASE BID AMOUNT	ADDENDUM NO.	
ALTERNATES (1)	UNIT PRICES (1)	UNIT PRICES (6)
(2)	(2)	(7)
(3)	(3)	(8)
(4)	(4)	(9)
(5)	(5)	(10)
SALES TAX	INSURANCE	

SCOPE OF WORK Including but not limited to the following:

This form must be completely filled out and signed by the subcontractor and the general contractor's representative to provide a record of this negotiation meeting.

ITEM	YES	NO	EXPLANATION AND/OR COMMENTS
1. Include all necessary labor, materials, and equipment as required to furnish and install all roofing and related sheet metal in accordance with the contract documents.			
2. When temporary utilities are available, they will be provided without cost. Where not available, subcontractor to provide at their cost.			
3. Subcontractor has reviewed spec. section(s) ____ and general and special conditions and agrees with all requirements pertaining to them.			
4. Insurance certificates per contract specifications will be provided. Hold-harmless clause, if required, to be included.			
5. Subcontractor has reviewed (company's) subcontract agreement and agrees to its terms.			
6. Flashing materials to be supplied to mason and/or carpenter - to be installed by their respective trades.			
7. Alternates, allowances, unit prices, if applicable, have been submitted and are attached hereto.			
8. Rubbish removal to occur on a daily basis and placed in dumpster located and provided by the general contractor.			
9. Hoisting of all materials and equipment to be provided by the subcontractor. Scaffolding, as required, to be supplied by the subcontractor.			

	FINAL AGREED AMOUNT
GENERAL CONTRACTOR'S REPRESENTATIVE SIGNATURE	SUBCONTRACTOR'S REPRESENTATIVE SIGNATURE

ORIGINAL

Figure 4.26 Subcontractor Form — Roofing - Sheet Metal. *(continued on next page)*

CONTINUATION SHEET

SUBCONTRACTOR NEGOTIATION FORM

Page 2 of 2

ITEM	YES	NO	EXPLANATION AND/OR COMMENTS
10. Subcontractor agrees to abide by all OSHA regulations and has read and accepts the general contractor's safety program.			
11. Additional work percentages: ____% profit, ____% overhead			
12. Before commencing work, subcontractor to inspect and accept all applicable substrates. If concrete, provide record of moisture test.			
13. Subcontractor will be responsible for removing rain/snow pertaining to their work.			
14. All vents and other roof penetrations to be sealed daily - either temporarily or permanently.			
15. Caulking of reglets and dissimilar roofing materials, as required, to be performmed by this subcontractor.			
16. Cleaning of any tar or roofing materials to be performed by this subcontractor and approved by the general contractor.			
17. Subcontractor to provide general contractor with written instructions on roof protection after roof has been completed and accepted.			
18. Payment and performance bonds to be furnished, if required by general contractor.			
19. All warranties to be in accordance with the contract specifications. Exceptions are as follows:			
20. All federal, state, local taxes are included. Taxes are not included, if tax exempt.			
21. Subcontractor is to return their subcontract agreement, if awarded, within ____ days of receipt. No work can commence until signed contract has been received and accepted by the general contractor.			
22. Other items to be addressed			

Figure 4.26 *(continued from previous page)* Subcontractor Form — Roofing - Sheet Metal.

SUBCONTRACTOR NEGOTIATION FORM

Page 1 of 2

PROJECT		
TRADE **Structural Steel**	SPECIFIED SECTIONS	DATE
SUBCONTRACTOR	REPRESENTED BY	(AREA CODE) TELEPHONE NO.
BASE BID AMOUNT	ADDENDUM NO.	
ALTERNATES (1)	UNIT PRICES (1)	UNIT PRICES (6)
(2)	(2)	(7)
(3)	(3)	(8)
(4)	(4)	(9)
(5)	(5)	(10)
SALES TAX	INSURANCE	

SCOPE OF WORK Including but not limited to the following:

This form is to be completely filled out and signed by the subcontractor and the general contractor's representative to provide a record of this negotiation meeting.

ITEM	YES	NO	EXPLANATION AND/OR COMMENTS
1. Furnish and install all structural steel in accordance with the plans and specs, including architectural and MEP drawings per attached list.			
2. Include all rolled sections, plates, bars, angles, tees, tubes, pipes, special shapes, and built-up members required to conform to all drawings.			
3. Shelf angles where called for on architectural and structural drawings.			
4. Quantity review- tonnage			
5. Erection - self or subcontracted, if so, to whom?			
6. Metal deck - breakout price, quantity. Furnish and install shear studs as required			
7. Includes miscellaneous iron per A,S.MEP requirements			
8. Fabrication (self) - location Fabricator - name and location			
9. Shop drawings- self? Other – specify ____________			

	FINAL AGREED AMOUNT
GENERAL CONTRACTOR'S REPRESENTATIVE SIGNATURE	SUBCONTRACTOR'S REPRESENTATIVE SIGNATURE

ORIGINAL

Figure 4.27 Subcontractor Form — Structural Steel. *(continued on next page)*

CONTINUATION SHEET

SUBCONTRACTOR NEGOTIATION FORM Page 2 of 2

ITEM	YES	NO	EXPLANATION AND/OR COMMENTS
10. Furnish all anchor bolts required for concrete to be set by others. Provide templates?			
11. Shop painting as required. Field touch up as required.			
12. Galvanized steel where called for			
13. Base plates, bearing plates, leveling nuts			
14. Concrete-filled columns			
15. Furnish and install masonry anchors, if required.			
16. Mill steel or warehouse steel?			
17. Erection bracing to be installed so as not to interfere with other trades.			
18. Provide temporary power, as required, to include electric welding machines.			
19. Hoisting of all materials, stairs, miscellaneous metal, and loose items.			
20. Subcontractor has received, read, and accepted contractor's safety program as well as all local, state, and federal (OSHA) safety regulations.			

Figure 4.27 *(continued from previous page)* Subcontractor Form — Structural Steel.

Section 5

Building Contractor's Comprehensive Checklist

SITEWORK – EARTHWORK INSPECTION CHECKS:	**Notes / Comments**
☐ Soil Investigative Report	
☐ Survey	
☐ Permits	
☐ Vegetation Protection	
☐ Existing Structure Protections	
☐ Existing Utility Lines	
☐ Shoring / Underpinning	
☐ Dust Control	
☐ Street Cleaning	
☐ Re-fueling Operations	
☐ Soil pile / Debris Locations	
☐ Grubbing / Soil Stripping	
☐ Soil Technician	
☐ Fencing	
☐ Temporary Road(s)	
☐ Drainage Erosion Control	
☐ Fill Material Approved	
☐ De-watering	
☐ Backfilling along new construction	
☐ Soil Poisoning	
☐ Record Drawings Updated	
☐ Erosion Control Plan	
☐ Hazardous Materials	
☐ Clean Up	
EARTHWORK – PILES / CAISSONS	
☐ Testing	
☐ Utilities Protected	
☐ Adjacent Structures Protected	
☐ Specifications Reviewed	
☐ Piles correct size and dimensions	
☐ Wood Piles treated	
☐ Waterproofing	
☐ Concrete Procedures / Testing	
☐ Reinforcing Steel	
☐ Records	
☐ Plumbness	
☐ Dewatering	

Sheet - 1 of 36

Figure 5.1 Building Contractor's Comprehensive Checklist. *(continued on next page)*

☐ Soils Technician	
☐ Outside Agency Inspections	
☐ Safety	
EARTHWORK – STORM DRAINAGE	
☐ Specifications	
☐ Excavations	
☐ Pipe Sizing	
☐ Slope / Gravity Flow	
☐ Tie Ins to Existing	
☐ Catch Basins – Manholes	
☐ Back fill	
☐ Soil Testing	
☐ Concrete Mix / Test	
☐ Grouting	
☐ Manufacturers Instructions	
☐ Local Codes	
☐ Workmanship	
☐ Safety	
☐ Clean Up	
☐	
EARTHWORK – PAVING AND SURFACING	
Subgrades	
☐ Specifications	
☐ Testing	
☐ Geo-Grid Soil Stablization	
☐ Subgrade to proper Elevation	
☐ Drains	
☐ Undergrounds- Utility	
☐ Equipment	
☐ Manholes, surface structure	
Priming and Tack Coats	
☐ Specifications	
☐ Inspection	
☐ Clean	
☐ Waterfree	
☐ Coverage	

Sheet - 2 of 36

Figure 5.1 *(continued from previous page)* Building Contractor's Comprehensive Checklist.

Asphalt Paving	
☐ Specifications	
☐ Adjacent Concrete Cured adequately (min 7 days)	
☐ Equipment	
☐ Correct Climate Specifications	
☐ Testing	
☐ Mix Design	
☐ Mix Temperature	
☐ Records	
☐ Truck Delivery Tags	
☐ Tie-ins	
☐ Correct Finish	
☐ Drainage	
☐ Walks, curbs, gutters , aprons to design	
☐ ADA Requirements	
☐ Surface level	
☐ Seal Coat	
☐ Striping	
☐ Disabled Parking	
☐ Clean Up	
Concrete Paving	
☐ Specifications	
☐ Equipment	
☐ Mix Design	
☐ Testing	
☐ Condition of Base Course	
☐ Forms / Bulkheads	
☐ Embeds	
☐ Joints	
☐ Reinforcement	
☐ Grade, slope, thickness controls	
☐ Correct Finish	
☐ Drainage	
☐ Walks,curbs, gutters , aprons to design	
☐ ADA Requirements	
☐ Clean up	

Sheet - 3 of 36

Figure 5.1 *(continued from previous page)* Building Contractor's Comprehensive Checklist.

CONCRETE – FORMING	
☐ Specifications	
☐ Type, dimensions, lumber grade	
☐ Formwork treatment	
☐ Weather Protection	
☐ Bracing, Construction	
☐ Spacers, Ties, Spreaders	
☐ Sleeves- Conduits, Piping	
☐ Block-outs	
☐ Embeds	
☐ Chamfer Strips	
☐ Cleanouts	
☐ Joints	
☐ Bulkheads	
☐ Alignments	
☐ Keyways	
☐ Openings (Vents, Doors....)	
☐ Shoring	
☐ Scaffolds / Equipment	
☐ Safety	
☐ Clean Up	
FORM REMOVAL	
☐ Curing Time	
☐ Damages	
☐ All Forming Removed / Fasteners	
☐ Reshoring	
☐ Patching	
☐ Re-treating	
☐ Safety	
☐ Clean Up	
CONCRETE – REINFORCEMENT	
☐ Specifications	
☐ Grades	
☐ Installation / Ties	
☐ Treatment	
☐ Clearances	

Sheet - 4 of 36

Figure 5.1 *(continued from previous page)* Building Contractor's Comprehensive Checklist.

☐ Splicing	
☐ Bends	
☐ Openings	
☐ Cleaned	
☐ Dowels	
☐ Supports	
☐ Secured	
☐ Sleeves	
☐ Catothotic Protection	
☐ Embeds	
☐ Agency Inspection	
☐ Safety	
☐ Clean Up	
CONCRETE – CAST IN PLACE	
☐ Specifications	
☐ Design Mix	
☐ Testing	
☐ Weather Conditions / Protection	
☐ Forming Acceptable	
☐ Equipment / Tools	
☐ Manpower	
☐ " Pockets " are vented	
☐ Joints	
☐ Concrete delivery	
☐ Embeds / Sleeves checked	
☐ Records	
☐ Rate of Pour	
☐ Safety	
☐ Clean Up	
FINISH AND CURING	
☐ Specifications	
☐ Correct Finish	
☐ Equipment	
☐ Curing Method	
☐ Joints	
☐ Embeds	
☐ Repairs	
☐ Weather Protection	

Sheet - 5 of 36

Figure 5.1 *(continued from previous page)* Building Contractor's Comprehensive Checklist.

☐ Loading	
☐ Safety	
☐ Clean Up	
PRE-CAST CONCRETE	
☐ Specifications	
☐ Inventory / delivery	
☐ Sample Comparisions	
☐ Special Items	
☐ Installers Qualified	
☐ Connectors	
☐ Installation	
☐ Equipment	
☐ Joints	
☐ Sealing	
☐ Flashings	
☐ Openings	
☐ Drip Edges	
☐ Alignment	
☐ Patching	
☐ Weather Protection	
☐ Safety	
☐ Clean Up	
MASONRY	
☐ Specifications	
☐ Material Storage	
☐ Mix Design	
☐ Testing	
☐ Sample Panel	
☐ Type , Size , Grade of Masonry Unit	
☐ Joints	
☐ Correct Pattern	
☐ Correct Cutting	
☐ Lintels / Bond Beams	
☐ Reinforcing	
☐ Sleeves	
☐ Embeds	

Sheet - 6 of 36

Figure 5.1 *(continued from previous page)* Building Contractor's Comprehensive Checklist.

☐ Flashing	
☐ Grouting	
☐ Clean-outs	
☐ Temporary / Permanent Supports	
☐ Weather Protection	
☐ Weep Holes	
☐ Hollow Metal Frames Installed Correctly	
☐ Moisture / Water Proofing	
☐ Repairs	
☐ Insulation	
☐ Back fill Protected	
CARPENTRY – ROUGH	
☐ Specifications	
☐ Storage	
☐ Tools / Equipment	
☐ Weather Protection	
☐ Grade	
☐ Defects	
☐ Treated	
☐ Alignment, Plumb, and Level	
☐ Connectors	
☐ Fasteners	
☐ Bracing	
☐ Notching, Drilling or Cutting	
☐ Bridging , blocking	
☐ Clearances	
☐ Supports , Bearing Areas	
☐ Spacing	
☐ Header Spans	
☐ Joist Spans	
☐ Rafter Spans	
☐ Outside Agency Inspection	
☐ Sheathing	
☐ Decking	
☐ Protection	
☐ Repairs	
☐ Correct Pitch	
☐ Safety	
☐ Clean Up	

Sheet - 7 of 36

Figure 5.1 *(continued from previous page)* Building Contractor's Comprehensive Checklist.

CARPENTRY – FINISH CHECKLIST	
☐ Specifications	
☐ Grade, Species , Type, Size	
☐ Condition of Items	
☐ Weather Protection	
☐ Sealed / Treatment / Finish	
☐ Stair Components / Tolerances	
☐ Workmanship	
☐ Fasteners	
☐ Joints	
☐ Alignment and Levelness	
☐ Doors and Drawers	
☐ Scribing Strips	
☐ Trim	
☐ Clearances Observed	
☐ Repairs	
☐ Safety	
☐ Clean-up	
METAL – STRUCTURAL STEEL	
☐ Specifications	
☐ Anchor Bolts	
☐ Deliveries checked	
☐ Weather Protection	
☐ Welds	
☐ Bolts / Shims	
☐ Storage	
☐ Treated / primed / Painted	
☐ Foundations / Piers clean	
☐ Temporary Bracing	
☐ Equipment	
☐ Safety	
☐ Clean Up	
Bolting	
☐ Specifications	
☐ Alignment	
☐ Special Washers	
☐ Testing	
☐ Tools Calibrated	

Sheet - 8 of 36

Figure 5.1 *(continued from previous page)* Building Contractor's Comprehensive Checklist.

☐ Free of Dirt (Paint ?)	
Welding	
☐ Specifications	
☐ Certified Personnel	
☐ Correct Weld	
☐ Testing	
☐ Protection	
☐ Safety	
☐ Clean Up	
METAL – FABRICATIONS	
☐ Specifications	
☐ Delivery and Storage	
☐ Bracing , Blocking, Anchoring	
☐ Protection	
☐ Templates (if needed)	
☐ Installation	
☐ Tolerances and Clearances	
☐ Welding	
☐ Sleeves , bolts, cutouts, holes, connectors	
☐ Stair treads and risers	
☐ Handrails	
☐ ADA Requirements	
☐ Finishes	
☐ Touch up paint	
☐ Code Requirements	
☐ Safety	
☐ Clean Up	
METAL – METAL JOISTS	
☐ Specifications	
☐ Unloading and Storage	
☐ Approved Coating	
☐ Equipment	
☐ Bridging and Anchoring	
☐ No cutting webs	
☐ Connections	

Sheet - 9 of 36

Figure 5.1 *(continued from previous page)* Building Contractor's Comprehensive Checklist.

☐ Welding	
☐ Alignment	
☐ Concentrated Loading	
☐ Rust, Scale slag cleaned	
☐ Safety	
☐ Clean up	
METAL – DECKING	
☐ Specifications	
☐ Approved Material	
☐ Storage	
☐ Accessory Items	
☐ Certified Welders	
☐ Provisions for Mech. , Elec Equipment	
☐ Decking Installation	
☐ Reinforcements	
☐ Embeds	
☐ Alignment	
☐ Adequate Supports	
☐ Seams	
☐ Concentric Loading	
☐ Shoring	
☐ Repairs	
☐ Penetrations	
☐ Paint Touch ups	
☐ Safety	
☐ Clean Up	
DOORS AND WINDOWS – METAL DOOR AND FRAMES	
☐ Specifications	
☐ Size, Type,Design, Finish	
☐ Weather Protected	
☐ Workmanship	
☐ Rated	
☐ Glazing	
☐ Straight, Level and Plumb	
☐ Swing	
☐ ADA Requirements	
☐ Hinge Reinforcing	

Sheet - 10 of 36

Figure 5.1 *(continued from previous page)* Building Contractor's Comprehensive Checklist.

☐ Hardware	
☐ Anchoring	
☐ Silencers	
☐ Weather Stripping / Thresholds	
☐ Special requirements	
☐ Repairs	
☐ Frames grouted	
☐ Safety	
☐ Clean Up	
DOORS AND WINDOWS – GLAZED CURTAIN WALLS	
☐ Specifications	
☐ Storage	
☐ Protection	
☐ Sizes, Shape and Thickness	
☐ Coatings	
☐ Sealants	
☐ Insulation	
☐ Dissimilar Metals	
☐ Weep Holes	
☐ Tolerances	
☐ Reveals	
☐ Anchorage	
☐ Fasteners	
☐ Workmanship	
☐ Safety	
☐ Clean Up	
DOORS AND WINDOWS - ENTNTRANCES AND STOREFRONTS	
☐ Specifications	
☐ Storage	
☐ Gauges, Patterns, colors	
☐ Protection	
☐ Coatings	
☐ Joints	
☐ Tolerances	
☐ Reveals	

Sheet - 11 of 36

Figure 5.1 *(continued from previous page)* Building Contractor's Comprehensive Checklist.

☐ Anchorage , Fasteners	
☐ Electric	
☐ Hardware	
☐ Door Swings	
☐ ADA Requirements	
☐ Safety	
☐ Clean Up	
DOORS AND WINDOWS – METAL WINDOWS	
☐ Specifications	
☐ Storage / Protection	
☐ Hardware	
☐ Special Items	
☐ Glazing Beads , Stops	
☐ Plumb, Square, Level	
☐ Protected Finish	
☐ Weathertight	
☐ Tolerance, clearance, operation	
☐ Rated	
☐ Glazing	
☐ Safety	
☐ Clean Up	
DOORS AND WINDOWS – WOOD AND PLASTIC DOORS	
☐ Specifications	
☐ Storage and Protection	
☐ Rough Opening	
☐ Door Swing	
☐ Jambs Blocking	
☐ Grade, Species, Size	
☐ Sealed	
☐ Tolerance / Clearances	
☐ Hardware	
☐ Lites / Glazing	
☐ Rating	
☐ ADA Requirements	
☐ Weather Stripping / Thresholds	

Sheet - 12 of 36

Figure 5.1 *(continued from previous page)* Building Contractor's Comprehensive Checklist.

☐ Repairs	
☐ Safety	
☐ Clean Up	
DOORS AND WINDOWS – HARDWARE	
☐ Specifications	
☐ Storage / protection	
☐ Manufacturer Templates	
☐ Finishes, Type, Style, Grade	
☐ Door Butts, Hinges	
☐ Closers	
☐ Stops	
☐ Special Hardware	
☐ Keying	
☐ Fire Door	
☐ Removed During Paint / Finishing	
☐ Safety	
☐ Clean Up	
Butts and Hinges	
☐ Specifications	
☐ Mortise	
☐ Hinge Placements	
☐ Sufficient Throw	
☐ Floor Hinges	
Locksets and Latchsets	
☐ Specifications	
☐ Drill , Boring	
☐ Backsets	
☐ Keys	
☐ Rated	
☐ Cylinder Cores	
☐ ADA	
☐ Workmanship	

Sheet - 13 of 36

Figure 5.1 *(continued from previous page)* Building Contractor's Comprehensive Checklist.

Door Closers	
☐ Specifications	
☐ Operation	
☐ Panic Device	
☐ ADA Requirements	
☐ Rating	
Stops , Holders and Plates	
☐ Facilitate Furniture	
☐ Blocking	
☐ Magnetic Holders	
☐ Push Plates	
☐ Kick Plates	
☐ Fasteners	
☐ Operation	
☐ Safety	
☐ Clean Up	
Miscellaneous Items	
☐ Sliding Door Hardware	
☐ Sliding Door Tracks	
☐ Weather Stripping	
☐ Thresholds	
☐ Fastening Devices	
☐ Safety	
☐ Clean Up	
DOORS AND WINDOWS – GLAZING	
☐ Specifications	
☐ Type, thickness, pattern, finish, labeled	
☐ Protection	
☐ Surface prep	
☐ Metal Primed	
☐ Rabbets and Beads	
☐ Clearance between glass and frames	
☐ Temperature Break	
☐ Rating	
☐ Fasteners	

Sheet - 14 of 36

Figure 5.1 *(continued from previous page)* Building Contractor's Comprehensive Checklist.

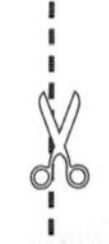

☐ Gaskets	
☐ Special Glazing	
☐ Safety Glass Locations	
☐ Treatments	
☐ Mirrors	
☐ Workmanship	
☐ Safety	
☐ Clean - Up	
THERMAL AND MOISTURE PROTECTION – WATERPROOFING	
☐ Specifications	
☐ Surfaces cleaned and prepped	
☐ Weather	
☐ Equipment / Tools	
☐ Penetrations	
☐ Joints	
☐ Corner details	
☐ Fasteners	
☐ Protected each day	
☐ Other Trade work	
☐ Backfilling (foundations)	
☐ Roofing Material Acceptable	
☐ Safety	
☐ Clean Up	
THERMAL AND MOISTURE PROTECTION – INSULATION	
RIGID BOARD INSULATION	
☐ Specifications	
☐ Correct R-Value	
☐ Weather Protection	
☐ Nailers	
☐ Vapor Barrier	
☐ Installation Method	
☐ Fasteners	
☐ Fire Proofing	
☐ Joints	

Sheet - 15 of 36

Figure 5.1 *(continued from previous page)* Building Contractor's Comprehensive Checklist.

☐ Slopes	
☐ Drains	
☐ Penetrations	
☐ Safety	
☐ Clean Up	
LOOSE FILL INSULATION	
☐ Specifications	
☐ Fire Rating	
☐ R- Value	
☐ Equipment	
☐ Installation Method	
☐ Safety	
☐ Clean Up	
THERMAL AND MOISTURE – FLASHING AND SHEET METAL	
☐ Specifications	
☐ Type, shape, gauge, metal	
☐ Finish / Treatment	
☐ Dissimilar Metals	
☐ Joints	
☐ Fasteners	
☐ Installation	
☐ Metal Lapping	
☐ Nailers / Cant Strips	
☐ Gravel Stops	
☐ Gutters	
☐ Scuppers	
☐ Accessories	
☐ Safety	
☐ Clean Up	
Downspouts	
☐ Slip Joints	
☐ Hangers / Supports	
☐ Color	
☐ Special Items	

Sheet - 16 of 36

Figure 5.1 *(continued from previous page)* Building Contractor's Comprehensive Checklist.

☐ Connections	
Base and Cap Flashing	
☐ Cant, Size, Gauge, Weight	
☐ Flange	
☐ Seams	
☐ Anchorage, Fasteners	
☐ Caps	
☐ Counter Flashing	
☐ Caulking	
Other Roof Flashing	
☐ Hip, Valleys and Ridges	
☐ Lapping	
☐ Fasteners	
☐ Reglets	
☐ Stepped Flashing	
☐ Caulking	
Through Wall Flashing	
☐ Fabrications	
☐ Laps	
☐ Locations	
☐ Sill and Pan Flashing	
Miscellaneous	
☐ Louvers	
☐ Screens	
☐ Shutters	
☐ Skylights, Hatches, Fans	
☐ Plastic Flashing	
☐ Termite Shields	
THERMAL AND MOISTURE PROTECTION – MEMBRANE ROOFING	
☐ Specifications	
☐ Storage	
☐ Surface Prep	

Sheet - 17 of 36

Figure 5.1 *(continued from previous page)* Building Contractor's Comprehensive Checklist.

☐ Nailers	
☐ Slope	
☐ Penetrations	
☐ Flashing	
☐ Fasteners	
☐ Temperature of Application	
☐ Weather Protected	
☐ Laps	
☐ Cant Strips, Reglets	
☐ Corner Details	
☐ Aggregates	
☐ Cap Sheets	
☐ Drains	
☐ Protected	
☐ Patching	
☐ Safety	
☐ Clean Up	
THERMAL AND MOISTURE PROTECTION – ROOFING SHINGLES TILES	
☐ Specifications	
☐ Grade, Type, Size, Rating, Pattern, Weight	
☐ Decking Prep	
Wood Shingles / Shakes	
☐ Underlayment	
☐ Weather	
☐ Fasteners	
☐ Exposure	
☐ Alignment	
☐ Valleys, Hips and Ridge	
☐ Flashing	
☐ Joints	
☐ Slope	
☐ Treated	
☐ Safety	
☐ Clean Up	

Sheet - 18 of 36

Figure 5.1 *(continued from previous page)* Building Contractor's Comprehensive Checklist.

Asphalt Shingles	
☐ Underlayment	
☐ Weather	
☐ Slope	
☐ Flashing	
☐ Alignment	
☐ Valleys, Hips, Ridge	
☐ Exposure	
☐ Fasteners	
☐ Safety	
☐ Clean Up	
Concrete / Clay Tiles	
☐ Underlayment	
☐ Exposure	
☐ Fasteners	
☐ Valleys, Hips and Ridge	
☐ Flashings	
☐ Joints	
☐ Slope	
☐ Treatment	
☐ Safety	
☐ Clean Up	
THERMAL AND MOISTURE PROTECTION- CEMENTITOUS FIREPROOFING	
☐ Specifications	
☐ Professional Applicators	
☐ Ratings , Thickness	
☐ Applied as per Manufacturer	
☐ Testing	
☐ Storage	
☐ Weather Protection	
☐ Application Temperature	
☐ Ventilation	
☐ Surface Preps	
☐ Hangers , Supports , Sleeves in place	
☐ Method of application	

Sheet - 19 of 36

Figure 5.1 *(continued from previous page)* Building Contractor's Comprehensive Checklist.

☐ Equipment	
☐ Curing Time	
☐ Other items covered / protected	
☐ Safety	
☐ Clean Up	
FINISHES – GYPSUM	
Framing	
☐ Specifications	
☐ Supporting System	
☐ Rating	
☐ Insulation	
☐ Anchorage	
☐ Fasteners	
☐ Hangers	
☐ Corner	
☐ Expansion	
☐ Casings, Trim	
☐ Access Panels	
☐ Blocking	
Wallboard	
☐ Specifications	
☐ Agency Inspections	
☐ Type, Thickness, Rating	
☐ Application	
☐ Weather Protected	
☐ Special Applications	
☐ Cut-outs	
☐ Joints	
☐ Fasteners	
☐ Taping	
☐ Coating	
☐ Drying Time	
☐ Sanding	
☐ Grade Finish	
☐ Corners	
☐ Safety	
☐ Clean - Up	

Sheet - 20 of 36

Figure 5.1 *(continued from previous page)* Building Contractor's Comprehensive Checklist.

FINISHES – LATH AND PLASTER	
☐ Specifications	
Framing and Furring	
☐ Stud Size, Type, Grade	
☐ Spacings	
☐ Locations, Layouts, and Plumbness	
☐ Channels aligned	
☐ Anchorage	
☐ Fasteners	
☐ Cut – Outs, Embeds	
☐ Hangers , Supports	
☐ Tie Wire	
☐ Elevations	
☐ Corner Beads	
☐ Expansion	
☐ Casings and Trim	
☐ Joints	
☐ Access Panels	
☐ Metal Lath , Type and Gauge	
☐ Sheathing	
☐ Agency Inspection	
☐ Safety	
☐ Clean Up	
Plaster	
☐ Specifications	
☐ Equipment	
☐ Application	
☐ Sensitive Items Covered	
☐ Temperature / Weather	
☐ Surface Prep	
☐ Color , Texture	
☐ Levelness	
☐ Safety	
☐ Clean - Up	

Sheet - 21 of 36

Figure 5.1 *(continued from previous page)* Building Contractor's Comprehensive Checklist.

FINISHES – PAINTING	
☐ Specifications	
☐ Storage	
☐ Color Schedules	
☐ Surface Prep	
☐ Dust Control	
☐ Temperature / Humidity	
☐ Lighting	
☐ Mixing	
☐ Primer	
☐ Number of Coats	
☐ Application Method	
☐ Workmanship	
☐ Paint Records	
☐ Curing / Drying Times	
☐ Corrections	
☐ Putty	
☐ Safety	
☐ Clean - Up	
FINISHES – ACOUSTICAL SYSTEMS	
Suspension System	
☐ Specifications	
☐ Storage	
☐ Weather Protection	
☐ Layout	
☐ Supports	
☐ Isolators	
☐ Sway Brackets	
☐ Alignment / Level	
☐ Rating	
☐ Access	
☐ Sound Proofing	
☐ Workmanship	
☐ Safety	
☐ Clean Up	
Acoustical Tile	
☐ Size, type, thickness, material, rating	

Sheet - 22 of 36

Figure 5.1 *(continued from previous page)* Building Contractor's Comprehensive Checklist.

☐ Weather Protected	
☐ Finish	
☐ Patching / Cutting	
☐ Uniform	
☐ Workmanship	
Acoustical Insulation	
☐ Batt are correct size, type, thickness, material, rating	
☐ Weather Protected	
☐ Secured	
☐ Hazardous Locations observed	
☐ Safety	
☐ Clean Up	
FINISHES – RESILIENT FLOORING	
☐ Specifications	
☐ Size, Type, Pattern, Weight, Color	
☐ Storage	
☐ Surface Prep	
☐ Primer	
☐ Cement Approved	
☐ Moisture Control	
☐ Tile Direction and Alignment	
☐ Sequence of Installation	
☐ Cement rate of Application (Thickness)	
☐ Workmanship	
☐ Corners and Stops	
☐ Joints	
☐ Levelness	
☐ Protection	
☐ Safety	
☐ Clean Up	
FINISHES - TILES	
☐ Specifications	
☐ Containers Sealed	
☐ Storage / Protection	
☐ Color, Size, Pattern, Shape, Type	

Sheet - 23 of 36

Figure 5.1 *(continued from previous page)* Building Contractor's Comprehensive Checklist.

☐ Trim Pieces	
☐ Grout Specs	
☐ Layout	
☐ Joints	
☐ Alignment	
☐ Equipment	
☐ Cuts	
☐ Bonding	
☐ Level	
☐ Sealed	
☐ Protected	
☐ Special Installation	
☐ Safety	
☐ Clean – Up	
Wall Tile – Mortar Set	
☐ Specifications	
☐ Mix	
☐ Support approved	
☐ Waterproofing	
☐ Embeds / Penetrations	
☐ Surface Prep	
☐ Bonding	
☐ Workmanship	
☐ Protection	
Wall Tile – Thin Set	
☐ Specifications	
☐ Setting Material	
☐ Support approved	
☐ Embeds / Penetrations	
☐ Surface Prep	
☐ Bonding	
☐ Workmanship	
☐ Protection	
Floor Tile	
☐ Specifications	
☐ Finish	
☐ Drains	

Sheet - 24 of 36

Figure 5.1 *(continued from previous page)* Building Contractor's Comprehensive Checklist.

☐ Slope	
☐ Joints	
☐ Waterproofing	
☐ Bonding	
☐ Level	
☐ Alignment	
☐ Sealed	
☐ Workmanship	
☐ Safety	
☐ Clean Up	
FINISHES – WOOD FLOORING	
Wood Strip Flooring	
☐ Specifications	
☐ Weather	
☐ Storage	
☐ Surface Prep	
☐ Moisture	
☐ Grade, Size, Type, and Species	
☐ Subflooring	
☐ Underlayment	
☐ Sleepers	
☐ Expansion	
☐ Joints	
☐ Fasteners	
☐ Pattern, Border, Color, Finish	
☐ Sanding	
☐ Workmanship	
☐ Safety	
☐ Protection	
☐ Repairs	
☐ Clean Up	
Parquet Flooring	
☐ Specifications	
☐ Weather	
☐ Storage	
☐ Surface Prep	
☐ Moisture Control / Vapor Barrier	
☐ Grade, Size, Type, and Species	

Sheet - 25 of 36

Figure 5.1 *(continued from previous page)* Building Contractor's Comprehensive Checklist.

☐ Subflooring	
☐ Underlayment	
☐ Sleepers	
☐ Expansion	
☐ Joints	
☐ Alignment	
☐ Fasteners	
☐ Pattern, Border, Color, Finish	
☐ Workmanship	
☐ Protection	
☐ Repairs	
☐ Safety	
☐ Clean Up	
FINISHES – CARPET	
☐ Specifications	
☐ Storage	
☐ Surface Prep	
☐ Padding	
☐ Tack Strips	
☐ Special Features	
☐ Seams	
☐ Direction of Installation	
☐ Cut outs, Stairs, Entries	
☐ Roll Sequence numbers	
☐ Workmanship	
☐ Protection	
☐ Safety	
☐ Clean-up	
MECHANICAL – AIR DISTRIBUTION SYSTEM	
☐ Specification	
☐ Approved Equipment w/nameplates	
☐ Mounting and Anchorage	
Furnaces	
☐ Specification	

Sheet - 26 of 36

Figure 5.1 *(continued from previous page)* Building Contractor's Comprehensive Checklist.

☐ Service Access	
☐ Fire Protection	
☐ Clearances	
Air Handling Units and Fans	
☐ Specification	
☐ Fan Rotation	
☐ Drive System	
☐ Guards	
☐ Protection	
☐ Roof Mounted	
☐ Supports	
☐ Dampers	
☐ Exhaust Air	
☐ Combustion Air	
☐ Testing	
☐ Safety	
☐ Clean –Up	
Filters and Screens	
☐ Specification	
☐ Size, Type	
☐ Accessible	
☐ Clean	
☐ Mounting	
☐ Sealing	
☐ Screens	
Ductwork	
☐ Specification	
☐ Layout	
☐ Type, Gauge, Shape	
☐ Joints	
☐ Slopes	
☐ Supports	
☐ Dampers	
☐ Fire Dampers	
☐ Access	
☐ Sleeves in walls and floors	

Sheet - 27 of 36

Figure 5.1 *(continued from previous page)* Building Contractor's Comprehensive Checklist.

☐ Sealed	
☐ Tested	
☐ Insulation	
☐ Fire Rating	
☐ Safety	
☐ Clean Up	
Outlets, Diffusers, Registers, and Grilles	
☐ Specification	
☐ Cleaned	
☐ Installation	
☐ Workmanship	
☐ Finishes	
☐ Controls	
☐ Supports	
☐ Balanced / Testing	
☐ Noise Control	
☐ Reports	
☐ Safety	
☐ Clean Up	
MECHANICAL – PLUMBING	
☐ Specifications	
☐ State , Local Approval (Stamped)	
☐ Grades / Layout	
☐ Foundation Sleeves	
☐ Pipe Size, Type, Grade	
☐ Supports	
☐ Expansion Requirements	
☐ Protection	
☐ Pipe Jointing	
☐ Welding	
☐ Valves, Unions and Fittings	
☐ Dielectric Protection	
☐ Coatings	
☐ Future Provisions	
☐ Testing	
☐ Agency Inspections	
☐ Safety	
☐ Clean Up	

Sheet - 28 of 36

Figure 5.1 *(continued from previous page)* Building Contractor's Comprehensive Checklist.

Soil, Waste and Vent	
☐ Specifications	
☐ Layout	
☐ Manholes, Cleanouts	
☐ Pipe Material, Type, Size, Grade	
☐ Dielectric Protection	
☐ Alignment and Slope	
☐ Supports	
☐ Drains	
☐ Traps	
☐ Fixture Rough Ins	
☐ ADA Requirements	
☐ Flashing	
☐ Testing	
☐ Agency Inspections	
☐ Safety	
☐ Clean – Up	
Water Supply Systems	
☐ Specifications	
☐ Layout	
☐ State , Local Approval	
☐ Pipe Size, Material, Grade	
☐ Joints	
☐ Shut off Valves	
☐ Dielectric Protection	
☐ Air Chambers	
☐ Expansion	
☐ Concealed Locations	
☐ Insulation	
☐ Sterilized	
☐ Agency Inspection	
☐ Safety	
☐ Clean Up	
Gas Piping System	
☐ Specifications	
☐ Layout	

Sheet - 29 of 36

Figure 5.1 *(continued from previous page)* Building Contractor's Comprehensive Checklist.

☐ State , Local Approval	
☐ Pipe Size, Material, Grade	
☐ Joints	
☐ Shut off Valves	
☐ Drip Legs	
☐ Dielectric Protection	
☐ Regulators	
☐ Protection	
☐ Agency Inspections	
☐ Safety	
☐ Clean Up	
Fixtures	
☐ Specifications	
☐ Blocking	
☐ Clearances	
☐ Installation	
☐ Protection	
☐ Special Fixtures	
☐ Vacuum Breakers	
☐ Temp / Pressure Valves	
☐ Drains	
☐ ADA Requirements	
☐ Testing	
☐ Safety	
☐ Clean Up	
MECHANICAL – BOILERS	
☐ Specifications	
☐ Manufacturers Instructions	
☐ Locations, Clearances	
☐ Name Plates	
☐ Fuel Burning	
☐ Controls	
☐ Venting	
☐ Storage Tanks	
☐ Expansion Tanks	
☐ Piping	
☐ Valves	
☐ Strainers, Gauges, Air Relief's, Drips,	

Sheet - 30 of 36

Figure 5.1 *(continued from previous page)* Building Contractor's Comprehensive Checklist.

Traps	
☐ Shut Offs	
☐ Temperature Relief Valves	
☐ Operational Controls	
☐ Combustionable Air	
☐ Exhaust System	
☐ Insulation	
☐ Testing	
☐ Agency Inspection	
☐ Balancing	
☐ Safety	
☐ Clean Up	
MECHANICAL – REFRIGERATION	
☐ Specification	
☐ Materials and Equipment Approved	
☐ Adequate Room	
☐ Mounting and Anchorage	
☐ Equipment Protection	
☐ Freeze Protection	
☐ Safety	
☐ Clean Up	
Piping	
☐ Specification	
☐ Type, Size, Grade	
☐ Fittings	
☐ Connectors	
☐ Slope	
☐ Supports	
☐ Air Vents	
☐ Balancing Cocks	
☐ Pressure Gauges	
☐ Insulation	
☐ Testing	
☐ Flashing / Sealing Through Wall	
☐ Condensating System	
☐ Safety	
☐ Clean Up	

Sheet - 31 of 36

Figure 5.1 *(continued from previous page)* Building Contractor's Comprehensive Checklist.

Equipment	
☐ Specification	
☐ Air Flow	
☐ Water Cooled Condensers	
☐ Evaporative Condensers	
☐ Reciprocating Compressor	
☐ Centrifugal Compressor	
☐ Noise and Vibration	
☐ Relief Valves	
☐ Shut –off Valves	
☐ Cooling Tower	
☐ Overflow and Drain Piping	
☐ Weather Protection	
☐ Pumps	
☐ Testing	
☐ Safety	
☐ Clean - Up	
MECHANICAL – FIRE PROTECTION	
☐ Specifications	
☐ State , Local Approval (Stamped)	
☐ Post Indicators	
☐ Siamese Connections	
☐ Layout	
☐ Expansion Joints	
☐ Test Stations	
☐ Service Interruptions	
☐ Piping	
☐ Piping Supports	
☐ Low Point Plugs	
☐ Heads	
☐ Protection	
☐ Drainage Valves	
☐ Wet System	
☐ Dry System	
☐ Power Requirements	
☐ Alarm Panel	
☐ Extinguishers / Cabinets	
☐ Alarm Stations	
☐ ADA Requirements	
☐ Safety	

Sheet - 32 of 36

Figure 5.1 *(continued from previous page)* Building Contractor's Comprehensive Checklist.

☐ Clean Up	
ELECTRIC – MATERIALS AND METHODS	
Raceways	
☐ Specifications	
☐ Approved Application Methods	
☐ Weather Protection	
☐ Conduit Size	
☐ Stub Ups	
☐ Layouts	
☐ Fittings	
☐ Dielectric Protection	
☐ Supports	
☐ Exposed Conditions	
☐ Coatings	
☐ Installation Depth	
☐ Pull Wires	
☐ Sleeves	
☐ Joints	
Busways	
☐ Specifications	
☐ Supports	
☐ Expansion Provisions	
☐ Grounded	
☐ Plug – Ins	
☐ Joints	
☐ Accessible	
Conductors	
☐ Specifications	
☐ Material, Size, Type	
☐ Pulling Operations	
☐ Connectors	
☐ Weather Protection	
☐ Junction and Outlet Boxes	
☐ Color Coding	
☐ Grounding	

Sheet - 33 of 36

Figure 5.1 *(continued from previous page)* Building Contractor's Comprehensive Checklist.

☐ Supports	
☐ Special Conditions	
Cable Systems	
☐ Specifications	
☐ Layouts	
☐ Protection	
☐ Supports	
☐ Special Locations	
☐ Inspections	
Outlets	
☐ Specifications	
☐ Layouts	
☐ Special Applications	
☐ Weather Protection	
☐ Floor Outlets	
☐ Accessible	
☐ Correct Rating	
☐ GFCI	
☐ Fire Protection	
☐ Sizes, Volumes Correct	
☐ Grounding	
☐ Plates	
☐ Supports	
☐ Safety	
☐ Clean Up	
Controls, Disconnects, and Starters	
☐ Specifications	
☐ Manual Disconnect	
☐ Locations	
☐ Nameplates, Data plates	
☐ Ratings	
☐ Dielectric connections	
☐ Special Installations	
☐ Manufacturer's Instructions	
☐ Testing	

Sheet - 34 of 36

Figure 5.1 *(continued from previous page)* Building Contractor's Comprehensive Checklist.

ELECTRIC – SERVICE AND DISTRIBUTION	
☐ Specifications	
☐ Utility Company Coordination	
☐ Clearances	
☐ Temporary Construction	
☐ Sleeves	
☐ Layout	
☐ Transformer	
Switchboards and Panelboards	
☐ Specifications	
☐ Temporary Electric	
☐ Mounting Height	
☐ Locations	
☐ Secured	
☐ Ground Fault Protection	
☐ Future Circuits	
☐ Spare Breakers	
☐ Working Space	
☐ Connections	
☐ Breakers Marked	
☐ Dielectric Protection	
☐ Grounding System	
☐ Weather Protection	
☐ Inspections	
☐ Safety	
☐ Clean Up	

Sheet - 35 of 36

Figure 5.1 *(continued from previous page)* Building Contractor's Comprehensive Checklist.

ELECTRIC – COMMUNICATION SYSTEMS	
☐ Specifications	
☐ Shop Drawings	
☐ Boxes, Conduits, Fittings	
☐ Low Voltage Protected	
☐ Grounded	
☐ Tested	
☐ Signal Devices Secure	
☐ Fire Alarm System	
☐ Security System	
☐ Visual Alarms	
☐ Local Fire Department	
☐ Agency Inspection	
☐ Safety	
☐ Clean Up	

Sheet - 36 of 36

Figure 5.1 *(continued from previous page)* Building Contractor's Comprehensive Checklist.

Section 6

Useful Forms for Builders

SUBCONTRACTOR LIST

Service	Vendor	Phone	Date

Figure 6.1 Example of a list for subcontractors.

Your Company Name
Your Company Address
Your Company Phone and Fax Numbers

LETTER SOLICITING BIDS FROM SUBCONTRACTORS

Date: ___________________
Subcontractor address: ___________________________
____
____

Dear: ______________________

I am soliciting bids for the work listed below, and I would like to offer you the opportunity to participate in the bidding. If you are interested in giving quoted prices for the LABOR / MATERIAL for this job, please let me hear from you. The job will start ____________. Financing has been arranged and the job will be started on schedule. Your quote, if you choose to enter one, must be received no later than ____________________.

The proposed work is as follows:
____
____
____
____

Thank you for your time and consideration in this request.

Sincerely,

Your Name
Title

Figure 6.2 Example of a letter soliciting bids from subcontractors.

CONTRACTOR QUESTIONNAIRE

PLEASE ANSWER ALL THE FOLLOWING QUESTIONS, AND EXPLAIN ANY "NO" ANSWERS.

Company name ____________________

Physical company address ____________________

Company mailing address ____________________

Company phone number ____________________

After hours phone number ____________________

Company President/Owner ____________________

President/Owner address ____________________

President/Owner phone number ____________________

How long has company been in business? ____________________

Name of insurance company ____________________

Insurance company phone number ____________________

Does company have liability insurance? ____________________

Amount of liability insurance coverage ____________________

Does company have Workman's Comp. insurance? ____________________

Type of work company is licensed to do ____________________

List Business or other license numbers ____________________

Where are licenses held? ____________________

If applicable, are all workman licensed? ____________________

Are there any lawsuits pending against the company? ____________________

Has the company ever been sued? ____________________

Does the company use subcontractors? ____________________

Is the company bonded? ____________________

Who is the company bonded with? ____________________

Has the company ever had complaints filed against it? ____________________

Are there any judgments against the company? ____________________

Please list 3 references of work similar to ours:

#1 ____________________

#2 ____________________

#3 ____________________

Please list 3 credit references:

#1 ____________________

#2 ____________________

#3 ____________________

Please list 3 trade references:

#1 ____________________

#2 ____________________

#3 ____________________

Please note any information you feel will influence our decision:

____

ALL OF THE ABOVE INFORMATION IS TRUE AND ACCURATE AS OF THIS DATE.

DATE:__________ COMPANY NAME: ____________________

BY:____________________ TITLE: ____________________

Figure 6.3 Example of a contractor questionnaire.

CONTRACTOR RATING SHEET

Job name: ________________ Date: ____________________________

Category	Contractor 1	Contractor 2	Contractor 3
Contractor name			
Returns calls			
Licensed			
Insured			
Bonded			
References			
Price			
Experience			
Years in business			
Work quality			
Availability			
Deposit required			
Detailed quote			
Personality			
Punctual			
Gut reaction			

Notes: ____

Figure 6.4 Example of a contractor rating form.

CONTRACTOR COMPARISON SHEET

Category	Contractor 1	Contractor 2	Contractor 3
CONTRACTOR NAME			
RETURNS CALLS			
LICENSED			
INSURED			
BONDED			
REFERENCES			
PRICE			
EXPERIENCE			
YEARS IN BUSINESS			
WORK QUALITY			
AVAILABILITY			
DEPOSIT REQUIRED			
DETAILED QUOTE			
PERSONALITY			
PUNCTUAL			
GUT REACTION			

Notes: ______________________________

Figure 6.5 Example of a contractor comparison form.

CONTRACTOR SELECTION FORM

TYPE OF SERVICE	VENDOR NAME	PHONE NUMBER	DATE SCHEDULED
Site Work	N/A		
Footings	N/A		
Concrete	N/A		
Foundation	N/A		
Waterproofing	N/A		
Masonry	N/A		
Framing	J. P. Buildal	231-8294	7/3/04
Roofing	N/A		
Siding	N/A		
Exterior Trim	N/A		
Gutters	N/A		
Pest Control	N/A		
Plumbing/R-I	TMG Plumbing, Inc.	242-1987	7/9/04
HVAC/R-I	Warming's HVAC	379-9071	7/15/04
Electrical/R-I	Bright Electric	257-2225	7/18/04
Central Vacuum	N/A		
Insulation	Allstar Insulators	242-4792	7/24/04
Drywall	Hank's Drywall	379-6638	7/29/04
Painter	J. C. Brush	247-8931	8/15/04
Wallpaper	N/A		
Tile	N/A		
Cabinets	N/A		
Countertops	N/A		
Interior Trim	The Final Touch Co.	365-1962	8/8/04
Floor Covering	Carpet Magicians	483-8724	8/19/04
Plumbing/Final	Same	Same	8/21/04
HVAC/Final	Same	Same	8/22/04
Electrical/Final	Same	Same	8/23/04
Cleaning	N/A		
Paving	N/A		
Landscaping	N/A		

NOTES/CHANGES ________________________________

____

Figure 6.6 Example of a contractor selection form.

SUBCONTRACTOR SCHEDULE

Type of Service	Vendor Name	Phone Number	Date Scheduled

Notes/Changes:

Figure 6.7 Example of a contractor schedule form.

Your Company Name
Your Company Address
Your Company Phone and Fax Numbers

INDEPENDENT CONTRACTOR ACKNOWLEDGMENT

Undersigned hereby enters into a certain arrangement or affiliation with Your Company Name, as of this date. The Undersigned confirms:

1. Undersigned is an independent contractor and is not an employee, agent, partner or joint venturer of or with the Company.
2. Undersigned shall not be entitled to participate in any vacation, medical or other fringe benefit or retirement program of the Company and shall not make claim of entitlement to any such employee program or benefit.
3. Undersigned shall be solely responsible for the payment of withholding taxes, FICA and other such tax deductions on any earnings or payments made, and the Company shall withhold no such payroll tax deductions from any payments due. The Undersigned agrees to indemnify and reimburse the Company from any claim or assessment by any taxing authority arising from this paragraph.
4. Undersigned and Company acknowledge that the Undersigned shall not be subject to the provisions of any personnel policy or rules and regulations applicable to employees, as the Undersigned shall fulfill his/her responsibility independent of and without supervisory control by the Company.

Signed under seal this ______ day of ____________ , 20___ .

____

____

Independent Contractor | Company Representative

____

Title

Figure 6.8 Example of an independent contractor acknowledgment form.

Your Company Name
Your Company Address
Your Company Phone and Fax Numbers

INDEPENDENT CONTRACTOR AGREEMENT

I understand that as an Independent Contractor I am solely responsible for my health, actions, taxes, insurance, transportation, and any other responsibilities that may be involved with the work I will be doing as an Independent Contractor.

I will not hold anyone else responsible for any claims or liabilities that may arise from this work or from any cause related to this work. I waive any rights I have or may have to hold anyone liable for any reason as a result of this work.

____

Independent Contractor Date

____

Witness Date

Figure 6.9 Example of an independent contractor agreement form.

Your Company Name
Your Company Address
Your Company Phone and Fax Number

CHANGE ORDER

This change order is an integral part of the contract dated ___________, between the customer _________________________, and the contractor, ____, for the work to be performed. The job location is _____________. The following changes are the only changes to be made. These changes shall now become a part of the original contract and may not be altered again without written authorization from all parties.

Changes to be as follows:

____

____

____

____

____

These changes will increase / decrease the original contract amount. Payment for theses changes will be made as follows: ____. The amount of change in the contract price will be ___________________________________ ($__________). The new total contract price shall be ________________________________ ($__________).

The undersigned parties hereby agree that these are the only changes to be made to the original contract. No verbal agreements will be valid. No further alterations will be allowed without additional written authorization, signed by all parties. This change order constitutes the entire agreement between the parties to alter the original contract.

______________________________	______________________________
Customer	Contractor
______________________________	______________________________
Date	Date

____
Customer

____
Date

Figure 6.10 Example of a change order form.

Your Company Name
Your Company Address
Your Company Phone and Fax Numbers

SUBCONTRACTOR QUESTIONNAIRE

Company name: ______________________________

Physical company address: ______________________________

Company mailing address: ______________________________

Company phone number: ______________________________

After-hours phone number: ______________________________

Company president/owner: ______________________________

President/owner address: ______________________________

President/owner phone number: ______________________________

How long has company been in business? ______________________________

Name of insurance company: ______________________________

Insurance company phone number: ______________________________

Does company have liability insurance? ______________________________

Amount of liability insurance coverage: ______________________________

Does company have worker's comp. insurance? ______________________________

Type of work company is licensed to do: ______________________________

List business or other license numbers: ______________________________

Where are licenses held? ______________________________

If applicable, are all workers licensed? ______________________________

Are there any lawsuits pending against the company? ______________________________

Has the company ever been sued? ______________________________

Does the company use subcontractors? ______________________________

Is the company bonded? ______________________________

With whom is the company bonded? ______________________________

Has the company had complaints filed against it? ______________________________

Are there any judgments against the company? ______________________________

Figure 6.11 Example of a subcontractor questionnaire.

Your Company Name
Your Company Address
Your Company Phone and Fax Numbers

SUBCONTRACTOR AGREEMENT

This agreement, made this __________ day of __________, 20__, shall set forth the whole agreement, in its entirety, between Contractor and Subcontractor.

Contractor: ____________________________________, referred to herein as Contractor.

Job location: ____________________________________

Subcontractor: ______________________________, referred to herein as Subcontractor.

The Contractor and Subcontractor agree to the following.

SCOPE OF WORK

Subcontractor shall perform all work as described below and provide all material to complete the work described below.

Subcontractor shall supply all labor and material to complete the work according to the attached plans and specifications. These attached plans and specifications have been initialed and signed by all parties. The work shall include, but is not limited to, the following: ____________________

____

____

____

____

COMMENCEMENT AND COMPLETION SCHEDULE

The work described above shall be started within _______ (____) days of verbal notice from Contractor, the projected start date is ___________. The Subcontractor shall complete the above work in a professional and expedient manner by no later than _______ (____) days from the start date. Time is of the essence in this contract. No extension of time will be valid without the Contractor's written consent. If Subcontractor does not complete the work in the time allowed, and if the lack of completion is not caused by the Contractor, the Subcontractor will be charged _______ ($______) dollars per day, for every day work extends beyond the completion date. This charge will be deducted from any payments due to the Subcontractor for work performed.

(Page 1 of 3. Please initial _________.)

Figure 6.12 Example of a subcontractor agreement. *(continued on next page)*

SUBCONTRACTOR AGREEMENT (continued)

CONTRACT SUM

The Contractor shall pay the Subcontractor for the performance of completed work subject to additions and deductions as authorized by this agreement or attached addendum. The contract sum is ____(\$__________).

PROGRESS PAYMENTS

The Contractor shall pay the Subcontractor installments as detailed below, once an acceptable insurance certificate has been filed by the Subcontractor with the Contractor. Contractor shall pay the Subcontractor as described: ____

All payments are subject to a site inspection and approval of work by the Contractor. Before final payment, the Subcontractor shall submit satisfactory evidence to the Contractor that no lien risk exists on the subject property.

WORKING CONDITIONS

Working hours will be _______ a.m. through _______ p.m., Monday through Friday. Subcontractor is required to clean work debris from the job site on a daily basis and leave the site in a clean and neat condition. Subcontractor shall be responsible for removal and disposal of all debris related to the job description.

CONTRACT ASSIGNMENT

Subcontractor shall not assign this contract or further subcontract the whole of this subcontract, without the written consent of the Contractor.

LAWS, PERMITS, FEES, AND NOTICES

Subcontractor shall be responsible for all required laws, permits, fees, or notices, required to perform the work stated herein.

WORK OF OTHERS

Subcontractor shall be responsible for any damage caused to existing conditions or other contractor's work. This damage will be repaired, and the Subcontractor charged for the expense and supervision of this work. The Subcontractor shall have the opportunity to quote a price for said repairs, but the Contractor is under no obligation to engage the Subcontractor to make said repairs. If a different subcontractor repairs the damage, the Subcontractor may be back charged for the cost of the repairs. Any repair costs will be deducted from any payments due to the Subcontractor. If no payments are due the Subcontractor, the Subcontractor shall pay the invoiced amount within _______ (_____) days.

(Page 2 of 3. Please initial ________.)

Figure 6.12 *(continued from previous page)* Example of a subcontractor agreement.

SUBCONTRACTOR AGREEMENT (continued)

WARRANTY

Subcontractor warrants to the Contractor, all work and materials for _________ from the final day of work performed.

INDEMNIFICATION

To the fullest extent allowed by law, the Subcontractor shall indemnify and hold harmless the Owner, the Contractor, and all of their agents and employees from and against all claims, damages, losses, and expenses.

This agreement, entered into on ________________, 20____, shall constitute the whole agreement between Contractor and Subcontractor.

______________________________ ______________________________
Contractor Date Subcontractor Date

(Page 3 of 3)

Your Company Name
Your Company Address
Your Company Phone and Fax Numbers

SUBCONTRACTOR CONTRACT ADDENDUM

This addendum is an integral part of the contract dated _______________, between the Contractor, ____________________, and the Customer(s), ______________________________, for the work being done on real estate commonly known as ___________.

The undersigned parties hereby agree to the following:

____
____
____
____
____

The above constitutes the only additions to the above-mentioned contract. No verbal agreements or other changes shall be valid unless made in writing and signed by all parties.

______________________________ ______________________________
Contractor Date Customer Date

____
Customer Date

Figure 6.12 *(continued from previous page)* Example of a subcontractor agreement.

Your Company Name
Your Company Address
Your Company Phone and Fax Numbers

LETTER OF ENGAGEMENT

Client: ____

Street: ____

City/State/Zip: ______________________________________

Work phone: ________________ Home phone: ________________

Services requested: __________________________________

____

____

Fee for services described above: $ ___________________

Payment to be made as follows: _______________________

____

By signing this letter of engagement, you indicate your understanding that this engagement letter constitutes a contractual agreement between us for the services set forth. This engagement does not include any services not specifically stated in this letter. Additional services, which you may request, will be subject to separate arrangements, to be set forth in writing.

A representative of ______________________ has advised us that we should seek legal counsel prior to using information or materials received from ______________________________________

____

We the undersigned hereby release ________________, its employees, officers, shareholders, and representatives from any liability. We understand that we shall have no rights, claims, or recourse and waive any claims or rights we may have against ________________, its employees, officers, shareholders, and representatives. We further understand that we will pay all costs of collection of any amount due hereunder including reasonable attorney's fees.

______________________________ ______________________________
Client Date Client Date

____
Company Representative Date

Figure 6.13 Example of a letter of engagement.